T0276850

Organic Pollutants: Concerns and Management

Organic Pollutants: Concerns and Management

Edited by **Bruce Horak**

New York

Published by Callisto Reference,
106 Park Avenue, Suite 200,
New York, NY 10016, USA
www.callistoreference.com

Organic Pollutants: Concerns and Management
Edited by Bruce Horak

International Standard Book Number: 978-1-63239-496-5 (Hardback)

Printed in the United States of America.

Contents

Preface

This book is a significant contribution as it investigates the impact of organic pollutants on human health and environment. Organic pollutants can lead to a number of environmental complications if discharged in air or water bodies. The existence of organic pollutants in the ecosystem, their risk as well as removal methodologies are issues of extreme significance. Various aspects of these pollutants have been considered in this book, with respect to their monitoring in soil and water, evaluation of their risks to humans, soil and plants, and the practical application of various strategies for their treatment and removal from the environment. Therefore, this book serves as an all-inclusive account on organic pollutants.

This book is the end result of constructive efforts and intensive research done by experts in this field. The aim of this book is to enlighten the readers with recent information in this area of research. The information provided in this profound book would serve as a valuable reference to students and researchers in this field.

At the end, I would like to thank all the authors for devoting their precious time and providing their valuable contribution to this book. I would also like to express my gratitude to my fellow colleagues who encouraged me throughout the process.

Editor

Section 1

Monitoring

Chapter 1

The Detection of Organic Pollutants at Trace Level by Variable Kinds of Silver Film with Novel Morphology

Zhengjun Zhang, Qin Zhou and Xian Zhang

Additional information is available at the end of the chapter

1. Introduction

In the modern world, environmental problems have attracted more and more attention, for environmental pollutants are extremely harmful to human beings' health. Environmental pollutants, such as persistent organic pollutants, are widely separated in the environment and difficult to detect at trace level.Within persistent organic pollutants, polychlorinated biphenyls (PCBs), due to their excellent dielectric properties, had been widely used since the 1920s in transformers, heat transfers, capacitors, etc., and had polluted nearly everywhere in the world [1]. In recent years, however, they have been found to be very harmful to human beings. They may cause serious diseases, such as cancers and gene distortion, when exceeding the critical dose in human bodies, and more seriously, PCBs can be accumulated in plants and animals from the environment and yield higher doses in human bodies, making PCBs very dangerous to human beings even in trace amounts [1-3]. Therefore, the detection of PCBs in trace amounts is crucial. Currently, the mostly applied detection technique for PCBs is the combination of high-resolution gas chromatography and mass spectrometry. It requires, however, very sophisticated devices, standard samples, complicated pretreatments of samples, favourable experimental environments and experienced operators [4-7]. Thus, new methods are demanded especially for the rapid detection of trace amounts of PCBs.

Surface-enhanced Raman scattering (SERS) has been proven to be an effective way to detect some organics [8]. With the great progress of nanoscale technology in recent years, SERS has attracted enormous attention due to its excellent performance and potential applications in the detection of molecules in trace amounts, even single molecule detection.

Among the approaches so far available to prepare nanostructure as SERS substrate, the glancing angle deposition (GLAD) technique is a simple but powerful means which is capable of producing thin films with pre-designed nanostructures. These nanostructures can be used in the field of SERS. For instance, using Ag nanorods as SERS substrates,Rhodamine 6G with concentration of 10^{-14} M (dissolved in water) was detected [9]; with the alumina-modified AgFON substrates, bacillus subtilis spores were detected to 10^{-14} M [10, 11]; Vo-Dinh reported even the detection of specific nucleic acid sequences by the SERS technique [12-14]. In spite of the numerous studies on the application as a chemical and biological sensor [15-17], the SERS technique has not yet been employed to detect PCBs as they are hardly dissolved in water.

2. Fabrication ofsilver nanostructure as sensitive SERS substrates

The detection sensitivity of SERS depends considerably on the surface property of the SERS substrate. High aspect ratio, nanostructured Ag, Au, Cu substrates are proved to be good SERS substrates. For instance, using ordered arrays of gold particles prepared through a porous alumina template as the SERS substrate, Rhodamine 6G (R6G) molecules were detected to a concentration limit of 10^{-12} M; arrays of silicon nanorods coated with thin films of Ag served as good SERS substrates for R6G molecule detection, etc [12, 16]. Thus the preparation of SERS substrates with preferred surface property is of great importance. There are several methods to prepare these kinds of SERS substrates and in this chapter we take glancing angle deposition as an example.

Glancing angle deposition (GLAD) technique is a simple but powerful means of producing thin films with pre-designed nanostructures, such as nanopillars, slanted posts, zigzag columns and spirals. Silver nanorod arrays prepared by GLAD are excellent SERS substrates.

In addition, the SERS properties are related to the optical properties of the nanorod arrays. Both SERS properties and optical properties depend on the structure of the nanorods, such as the shape, length, separation, tilting angle and so on, which can be tuned by the deposition conditions.

2.1. Fabrication of sensitive SERS substrates by GLAD

The detection sensitivity of the SERS technique depends greatly on the surface property of the SERS substrate [45,46]. Among the approaches so far available to prepare nanostructured materials, the glancing angle deposition (GLAD) technique is a simple but powerful means of producing thin films with pre-designed nanostructures [47-48], such as nanopillars, slanted posts, zigzag columns, spirals [18-19], etc [20-23]. For example, arrays of Ag nanorods were found to be good SERS substrates for the detection of trans-1,2-bis(4-pyridyl)ethane molecules, with a SERS enhancement factor greater than 10^8 [16]. It is therefore of great interest to investigate the growth of metal nanostructures by the GLAD technique [49].

Pristine Si wafers with (001) orientation were used as substrates. These were supersonically cleaned in acetone, ethanol and de-ionized water baths in sequence, and were fixed on the GLAD substrate in an e-beam deposition system. The system was pumped down to a vacuum level of 3×10^{-5} Pa and then the thin Ag film was deposited on the substrate with a depositing rate of 0.5 nm/s, with the thickness monitored by a quartz crystal microbalance. To produce films of aligned Ag nanorods, the incident beam of Ag flux was set at ~ 85 ° from the normal of the silicon substrate, at different substrate temperatures. The morphology and structure of the thin Ag films was characterized by scanning electron microscope (SEM), transmission electron microscope (TEM) and high-resolution TEM, selected area diffraction (SAD) and X-ray diffraction (XRD), respectively. The performance of the nanostructured Ag films as SERS substrates was evaluated with a micro-Raman spectrometer using R6G as the model molecule.

It is well known that the major factors influencing the growth morphology of the films by GLAD are the incident direction of the depositing beam flux, the temperature and the movement of the substrate, and the deposition rate, etc. When fixing the incident Ag flux at ~85 ° from the normal of the substrate and the deposition rate at ~0.5 nm/s, the growth morphology of the Ag films was greatly dependent on the temperature and movement of the substrate. Fig 1 shows the growth morphology of thin Ag films versus the temperature and movement of the substrate. The SEM micrographs were taken by a FEI SEM (QUANTA 200FEG) working at 20 kV.

Fig 1(a) and (b) shows typical SEM images of the surface morphology of thin Ag films deposited at 120 °C, without substrate rotation and with substrate rotation at a speed of 0.2 rpm, respectively. One sees from the images that at this temperature, Ag nanorods formed in two films with a length of 500 nm, yet they were not well separated - most nanorods were joined together. A major difference between the two is the growth direction of the joined nanorods, i.e. without rotation the nanorods grew at a glancing angle on the substrate, while with substrate rotation the nanorods grew vertically aligned. Another difference noticeable is the size of the nanorods, i.e. nanorods grown with substrate rotation have a slightly larger diameter.

Fig 1(c) and (d) shows respectively the surface morphology of thin Ag films deposited at -40 °C, without substrate rotation and with rotation at a speed of 0.2 rpm. Comparing with Figs 1(a) and (b), it can be seen that the decrease in the deposition temperature led to the separation of Ag nanorods in the two films, while the rotation of the substrate also determined the growth direction and diameter of the nanorods, as observed from Figs 1(a) and 1(b). The Ag nanorods grown at this temperature are 20-30 nm in diameter, ~ 800 nm in length and are well separated. Therefore, through adjusting the temperature and movement of the substrate one can grow well separated and aligned Ag nanorods on planar silicon substrates.

Fig 1(e) and 1(f) shows respectively a bright-field TEM and a HRTEM image of Ag nanorods shown by Fig 1(c); inset of Fig 1(f) is the corresponding SAD pattern. The images and the SAD pattern were taken with a JEM-2011F working at 200 kV. One sees from the Figs that the Ag nanorod is ~ 30 nm in diameter and its micro-structure is single crystalline. By indexing the SAD pattern it is noticed that during the growth process the {111} plane of the nanorod was parallel to the substrate surface, with its axis along the <110> direction. This was confirmed by XRD analysis. Fig 2 shows a XRD pattern of the Ag nanorods shown by Fig 1(c). The pattern was taken with a Rigaku X-ray diffractometer using the Cu k. line, working at the θ-2θ coupled scan mode. From the Fig, a very strong (111) texture is observed, indicating that the {111} plane of the Ag nanorods was parallel to the substrate surface. These suggest that one can produce arrays of aligned, single crystalline Ag nanorods by the GLAD technique even at a low substrate temperature, i.e. -40 °C.

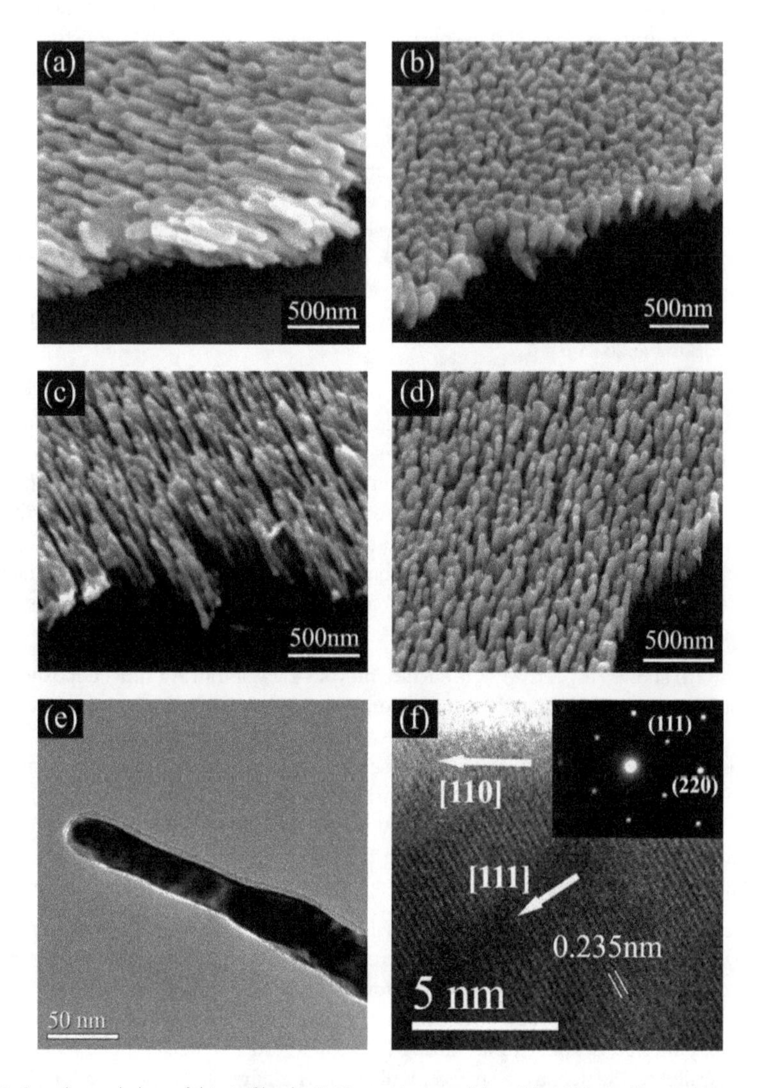

Figure 1. Growth morphology of thin Ag films by GLAD at various conditions. (a) at 120 °C without substrate rotation; (b) at 120 °C and substrate rotation at 0.2 rpm; (c) at -40 °C without substrate rotation; and (d) at -40 °C and substrate rotation at 0.2 rpm. (e) and (f) shows respectively a bright-field TEM and a HRTEM image of the nanorods shown by Fig 1(c); inset of (f) is the corresponding SAD pattern.

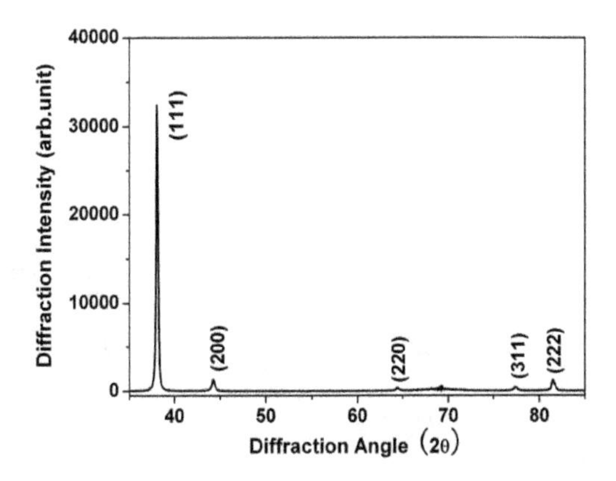

Figure 2. A XRD pattern of the Ag film consisting of well separated, single crystalline nanorods shown by Fig 1(c).

By using Rhodamine 6G as the model molecule, the performance of thin Ag films shown by Figs 1(a)-(d) is examined as the SERS substrates. These samples were dipped in a 1×10^{-6}mol/L solution of R6G in water for 30 minutes and dried with a continuous gentle nitrogen blow. Fig 3(a) and 3(b) show Raman spectra of R6G obtained on the four nanostructured Ag films by a Reinshaw 100 Raman spectrometer using a 514 nm Ar^+ laser as the excitation source. It is observed that with the thin Ag films as the SERS substrate, all spectra exhibit clearly the characteristic peaks of R6G molecules, at 612, 774, 1180, 1311, 1361, 1511, 1575 and 1648 cm^{-1}, respectively12.However, the intensity of the Raman peaks was dependent on the morphology of the films. It is noticed that on Ag films consisting of well separated nanorods, see Figs 3(b), the Raman peaks of R6G are much stronger than those on films of joined nanorods, see Figs 3(a). This suggests that arrays of aligned but well separated Ag nanorods represent excellent SERS performance.

Using arrays of aligned Ag nanorods shown by Figs 1(c) and 1(d) as SERS substrates, we examined the detection limit of R6G molecules in water by the SERS technique. Fig 4(a) shows Raman spectra of R6G obtained on Ag nanorods shown by Fig 1(c), as a function of the concentration of R6G in water ranging from 1×10^{-8} to 1×10^{-16}mol/L. Similar results were also obtained for Ag nanorods shown by Fig 1(d). The Raman spectra were obtained by one scan with an accumulation time of 10 s, at a laser power of 1 % to avoid decomposition of R6G. It is found that characteristic peaks of R6G were all observed at all concentrations. To clearly show this, we plot the Raman spectrum at 10^{-14}mol/L in Fig 4(b). It is noticed that although the intensity of the peaks is almost two orders lower than that at 10^{-6}mol/L, the spectrum contains the clear characteristic peaks of R6G12. These suggest that Ag films consisting of aligned and well separated Ag nanorods with single crystalline could serve as excellent SERS substrate for trace amount detection of R6G molecules. However, in the Raman spectrum at 10^{-16}mol/L in Fig 4(a), some of the peaks of R6G disappear. That suggests the concentration limit of this method is 10^{-14}mol/L in the authors' work.

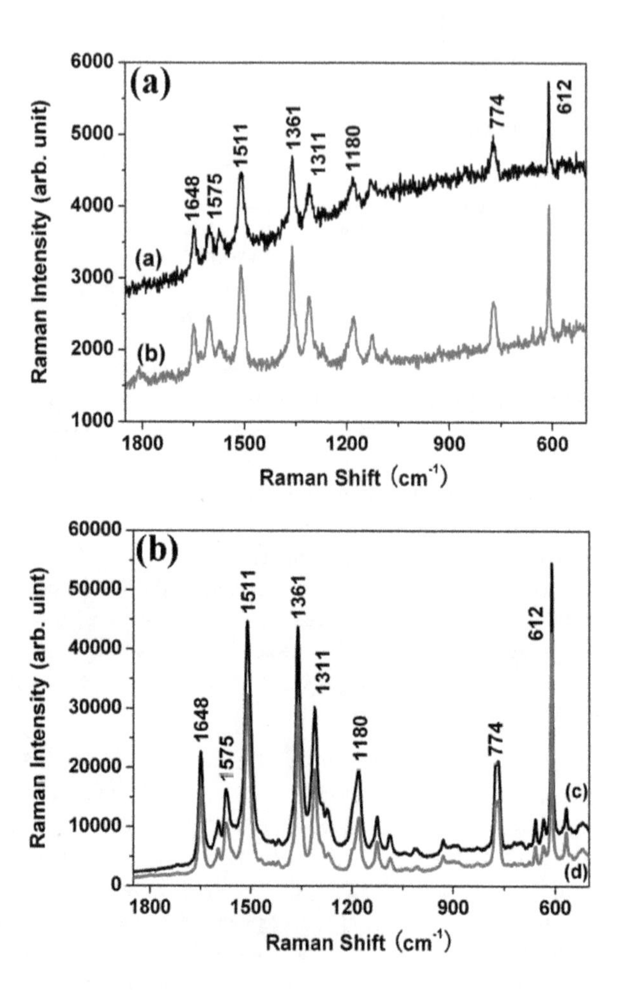

Figure 3. Raman spectra of R6G on thin Ag films consisting of (a) joined nanorods shown by Figs 1(a) (black line) and 1(b) (grey line); and (b) separated Ag nanorods shown by Figs 1(c) (black line) and 1(d) (grey line), respectively, at a concentration of 1×10^{-6}mol/L.

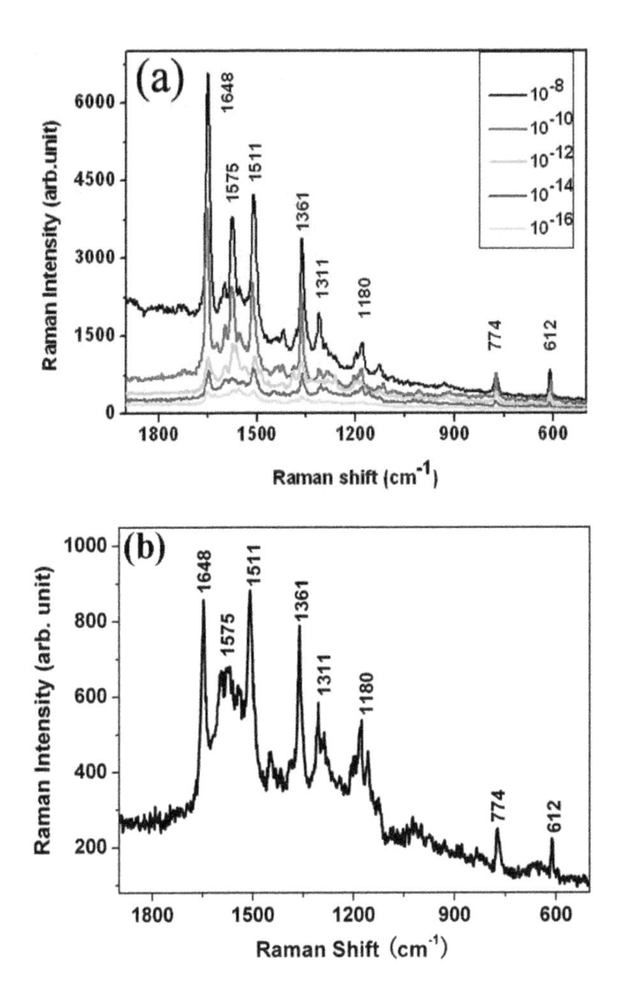

Figure 4. a) Raman spectra of R6G at concentrations ranging from 1×10^{-8} to 1×10^{-16}mol/L; and (b) the Raman spectrum of R6G at a concentration of 1×10^{-14}mol/L, on the thin Ag film consisting of well separated, single crystalline Ag nanorods.

2.2. Enhancement of the sensitivity of SERS substrates via underlayer films

Although the Ag nanorod arrays present sensitive SERS performance, it is still necessary to enable the substrate to detect organic pollutants at trace amount with adequate sensitivity. There are several ways to promote the sensitivity of Ag nanorods as SERS substrates.

Much effort has been devoted to achieving highly sensitive SERS substrates. In particular, multilayer structures can improve SERS enhancement, such as "sandwich" structures with silver oxide or carbon inside and Ag or Au as both underlayer and overlayer [24-28]. Other researchers found that multilayer structures of Ag/Au nanostructures on the smooth metallic underlayer exhibited better SERS sensitivity compared to those without metallic underlayer ($EF = 5 \times 10^8$) [29-31]. However, the factor that governs the enhancement for multilayer structures is not very clear. Recently, Misra*et al.* obtained remarkably high SERS sensitivity using a micro-cavity with a radius of several micrometers [32]. Shoute*et al.* obtained high SERS signals ($EF = 6 \times 10^6$) for molecules adsorbed on the silver island films supported by thermally oxidized silicon wafers and declared that the additional enhancement was due to the optical interference effect [33]. All the above experiments and those conducted by Driskell*et al.* suggested that the underlayer reflectivity could play an important role in the multilayer SERS substrates [29].

We have investigated in detail the relationship of underlayer reflectivity and the SERS enhancement of Ag nanorod substrates prepared by oblique angle deposition. We use thin films of different materials with different thicknesses as underlayers to modulate the reflectivity systematically. With the coating of the same Ag nanorods, we find that the SERS intensity increases linearly with the underlayer reflectivity. This conclusion can be explained by a modified Greenler's model we recently developed [34].

To change the reflectivity of the underlayer films, one can vary the dielectric constant and the thickness of the films systematically. We proposed to use Ag, Al, Si and Ti films, since they have different dielectric constants and can be fabricated easily. With a transfer matrix method, we can calculate the reflectivity of those films [35, 36].Fig 5(a) shows the calculated reflectivity spectra of 100 nm Ag, Al, Si, and Ti films. In general, the reflectivity, $R_{Ag}>R_{Al}>R_{Ti}>R_{Si}$, except that at $\lambda \sim 600$ nm where the Si film has a large constructive interference. Fig 5(b) plots the film thickness dependent reflectivity for Ag, Al, Si and Ti at a fixed wavelength $\lambda_0 = 785$ nm. The reflectivity of Ag, Al, Ti, e.g. metals, increases monotonically with the film thickness d. The reflectivity R of Ag, Al and Ti thin films increases sharply when $d<$ 100 nm and almost remains unchanged when 100 nm $\leq d \leq$ 400 nm, while R_{Si} shows an oscillative behaviour due to the interference effect of a dielectric layer.

We deposited thin Ag, Al, Si and Ti films, all with thickness $d = 25$, 100, and 400 nm, respectively, to achieve different reflectivity. All depositions were carried out in a custom-designed electron-beam deposition system [16].Before the deposition, the glass slide substrates were cleaned by piranha solution ($H_2SO_4 : H_2O_2 = 4:1$ in volume). The pellets of source materials, Ag, Al, Ti, with 99.99% purity, were purchased from Kurt J. Lesker Company, and Si with 99.9999% purity was purchased from Alfa Aesar Company. The film thickness was monitored *in situ* by a quartz crystal microbalance (QCM) facing toward the vapour source. After the deposition, the reflectivity of the deposited thin films was measured by an Ultraviolet-Visible Spectrophotometer (UV-Vis) double beam spectrophotometer with an integrating sphere (Shimadzu UV-Vis 2450). Fig 5(c) shows the reflectivity spectra of the twelve thin films obtained. The shapes of the reflection spectra are qualitatively consistent with those predicted by the calculations, as shown in Figs. 5(a) and (b). At the same wavelength, in general, $R_{Ag}>R_{Al}>R_{Ti}>R_{Si}$. In the visible wavelength region, the reflectivity of Ag, Al and Ti increases with the thickness d, while Si demonstrates an oscillating behaviour.

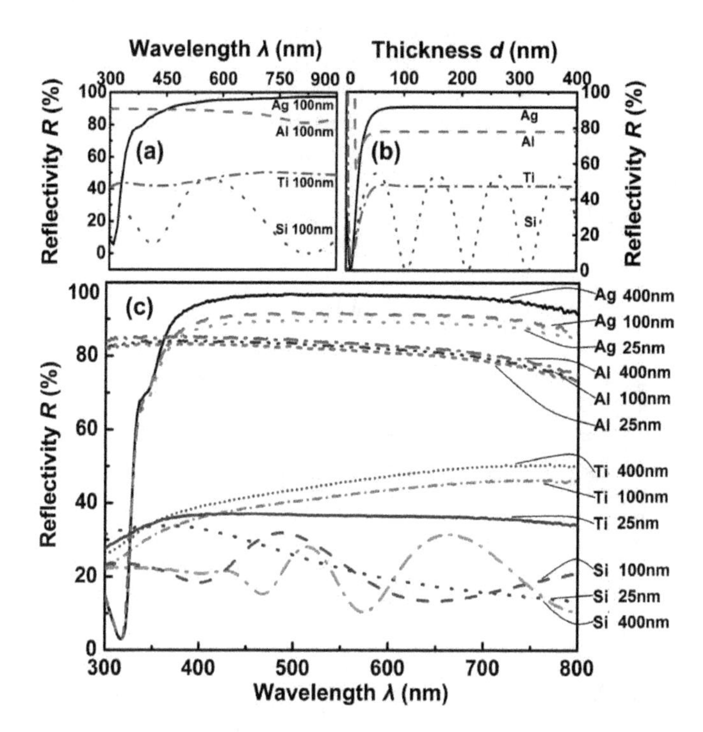

Figure 5. a) Calculated reflectivity R of thin Ag, Al, Si and Ti films at different wavelengths λ with film thickness of 100 nm; b) calculated reflectivity R of thin Ag, Al, Si and Ti films with different thicknesses d at λ_0 = 785 nm; c) experimentally obtained reflectivity spectra of thin Ag, Al, Si and Ti films with different thicknesses.

The twelve deposited planar thin film samples were then loaded into another custom-designed electron-beam evaporation system for Ag nanorod deposition through the so-called oblique angle deposition (OAD) [16, 29, 37].In this deposition, the background pressure was 1×10^{-7}Torr and the substrate holder was rotated so that the deposition flux was incident onto the thin films with an angle $\theta = 86^{\circ}$ with respect to the surface normal of the substrate holder. The Ag nanorod arrays were formed through a self-shadowing effect [16, 29, 37]. During the deposition, the Ag deposition rate was monitored by a QCM directly facing the incident vapour. The deposition rate was fixed at 0.3 nm/s and the deposition ended when the QCM read 2000 nm (our optimized condition).

The morphologies of the Ag nanorod arrays on different thin film substrates were characterized by a scanning electron microscope (SEM, FEI Inspect F). The typical top-view SEM images are shown in Fig. 6 and they all look very similar. From the cross-section and top-view SEM images, the length *L*, diameter *D* and separation *S* of these Ag nanorods on different planar thin films are obtained statistically: L_{Ag}= 940 ± 70 nm, D_{Ag}= 90 ± 10 nm, S_{Ag}= 140 ± 30 nm; L_{Al}= 950 ± 50 nm, D_{Al}= 90 ± 10 nm, S_{Al}= 140 ± 30 nm; L_{Si}= 900 ± 50 nm, D_{Si}= 80 ± 10 nm, S_{Si}= 130 ± 20 nm; and L_{Ti}= 930 ± 60 nm, D_{Ti}= 90 ± 10 nm, S_{Ti}= 130 ± 20 nm, respectively. The Ag nanorod tilting angles β were measured to be about 73ºwith respect to substrate normal, which are consistent with our previous results [16, 29, 37]. These structure parameters are very close to one another, implying that the Ag nanorod arrays deposited on different thin film substrates are statistically the same. The SERS response of these Ag nanorod substrates were evaluated under identical conditions: A 2 μL droplet of a Raman probe molecule, trans-1, 2- bis (4-pyridyl) ethylene (BPE) with a concentration of 10^{-5} M, was uniformly dispersed onto the Ag nanorod substrates. The SERS spectra were recorded by the HRC-10HT Raman Analyzer from EnwaveOptronics Inc., with an excitation wavelength of λ_0 = 785 nm, a power of 30 mW and an accumulation time of 10 s.

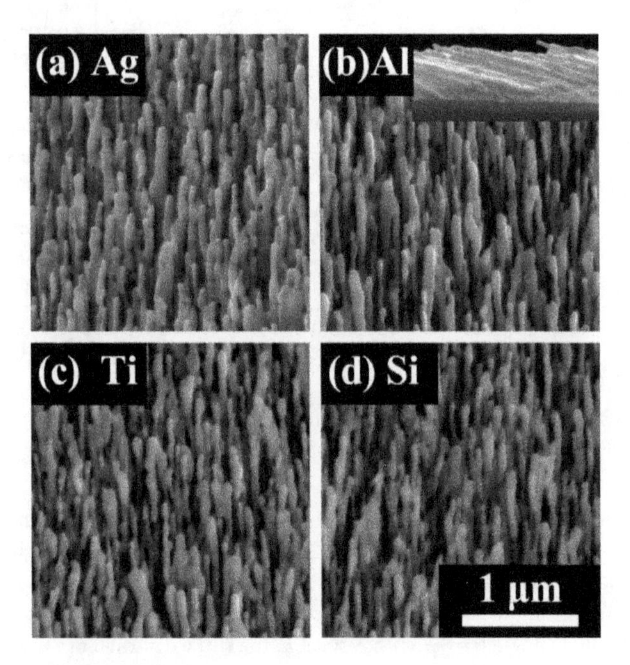

Figure 6. Representative SEM images of Ag nanorod arrays on 100 nm underlayer thin films with different materials: (a) Ag; (b) Al; (c) Ti; (d) Si. All the Figs have the same scale bar.

Fig 7(a) shows the representative BPE SERS spectra obtained at λ_0 = 785 nm from the Ag nanorod arrays on thin Ag, Al, Si and Ti film underlayers (thickness *d* = 100 nm). Each spectrum is an average of at least 15 different spectra taken at different spots on the substrates. All of them show the three main Raman bands of BPE, $\Delta\nu$ = 1639, 1610, and 1200 cm^{-1}, which can be assigned to the C=C stretching mode, aromatic ring stretching mode and in-plane ring mode, respectively [38]. The SERS intensity of the Ag nanorods grown on thin Ag film are higher than others and the SERS intensity of the Ag nanorods on Al film is larger than that on Ti film. The Ag nanorods on Si film show the smallest SERS intensity. According to Fig. 5(c), this seems to follow a trend: the larger the underlayer reflectivity, the larger the SERS intensity. To quantitatively compare the SERS response of these substrates, the Raman peak intensity I_{1200} at $\Delta\nu = 1200$ cm^{-1} is analyzed.

Fig 7(b) plots the SERS intensity I_{1200} versus the reflectivity R of the underlayer thin films at $\lambda_0 = 785$ nm. The error bar for the Raman intensity is the standard deviation from 15 or more measurements from multiple sampling spots on the same substrates and the error bar for the reflectivity data is calculated from multiple reflectivity measurements at $\lambda_0 = 785$ nm. In Fig. 7(b), the SERS intensity and reflectivity follow a linear relationship: when the reflectivity of the underlayer increases, the SERS enhancement factor increases.

This linear relationship of the underlayer reflectivity and SERS intensity can be explained by a modified Greenler's model developed by Liu *et al.* [34]Greenler's model is proposed through classical electrodynamics to explain the effects of the incident angles and polarization, and the collecting angle on the Raman scattering from a molecule adsorbed on a planar surface [32]. The modified Greenler's model extended the Greenler's model from a planar surface to Ag nanorod substrates and considered the effect of the underlying substrate [34]. The main point of the modified Greenler's model is to consider the conditions of both the incident and scattering fields near the molecule absorbed on a nanorod to calculate the enhancement.

Assuming that relative Raman intensity η is the ratio of the total Raman scattering power to incident light power, according to the modified Greenler's model, η excited by an unpolarized light can be explicitly expressed as [34]

$$\eta = \langle E_{Raman}^2 \rangle / \langle E_{incident}^2 \rangle$$
$$= \frac{1}{2}\Big\{[1 + R_p + n_2^4 R'_p \cos\delta_p \cos 2(\varphi-\beta) + 2n_2^2 R'^{\frac{1}{2}}_p \cos(\delta'_p + 2\pi\Delta/\lambda)\sin 2\beta$$
$$+2n_2^2 R_p^{\frac{1}{2}} R'^{\frac{1}{2}}_p \sin 2\varphi \cos(\delta'_p + 2\pi\Delta/\lambda - \delta_p)](1 + R_p + 2\sqrt{R_p}\cos\delta_p)\cos^2(\varphi-\beta)$$
$$+[1 + n_2^4 R'_s + 2n_2^2 R'^{\frac{1}{2}}_s \cos(\delta'_s + 2\pi\Delta/\lambda)]\Big\}$$

where R_p and R_s are the reflectivity of *p*- and *s*-polarized lights by the Ag nanorod surface, and R'_p and R'_s are the reflectivity of *p*- and *s*-polarized components by the underlayer thin film; n_2 is complex refractive index of Ag, and $n_2 = 0.03 + 5.242i$ (for $\lambda_0 = 785$ nm); φ is the light incident angle, and β is the Ag nanorod tilting angle; $\Delta = d(1 + \cos 2\varphi)/\cos\varphi$, where d is the thickness of Ag nanorod layer; δ_p, δ_s, δ'_p, and δ'_s are the reflectivity phase shifts of *p*- and *s*-polarization E-fields from Ag nanorods and underlayer thin film, defined as

$$\delta_p = \tan^{-1}[\mathrm{Im}(r_p)/\mathrm{Re}(r_p)],\ \delta_s = \tan^{-1}[\mathrm{Im}(r_s)/\mathrm{Re}(r_s)]$$
$$\delta'_p = \tan^{-1}[\mathrm{Im}(r'_p)/\mathrm{Re}(r'_p)],\ \delta'_s = \tan^{-1}[\mathrm{Im}(r'_s)/\mathrm{Re}(r'_s)]$$

By setting the light incident angle $\varphi = 0^{\circ}$, the Ag nanorod tilting angle $\beta = 73^{\circ}$, the thickness of Ag layer $d = 300$ nm, the relative Raman intensity η as a function of the underlayer reflectivity R at $\lambda_0 = 785$ nm is calculated and plotted in Fig. 7(c). It shows that the η indeed increases linearly with R, which is in very good agreement with our experimental data shown in Fig. 7(b). Therefore, the underlayer reflectivity is one significant parameter to consider for improving the SERS response of multilayer substrates.

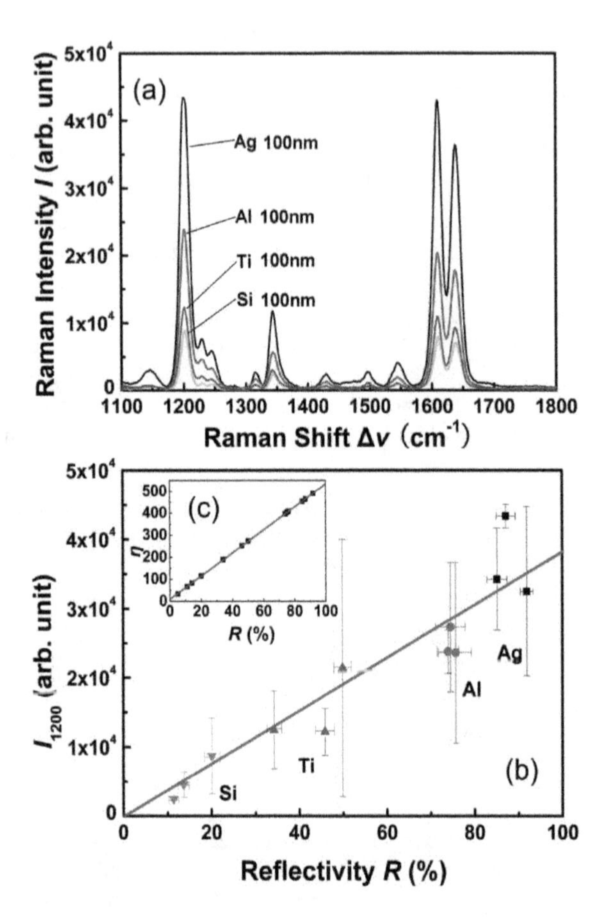

Figure 7. a) BPE SERS spectra obtained from Ag nanorod arrays deposited on 100 nm thin Ag, Al, Si and Ti film underlayers; (b) the plot of experimental Raman intensity as a function of underlayer reflectivity at $\lambda_0 = 785$ nm. Different symbol groups represent different kinds of substrates. (c) The plot of the enhanced Raman intensity ratio η as a function of underlayer reflectivity calculated by the modified Greenler's model.

Both our experiments and the modified Greenler's model demonstrate that the higher the underlayer reflectivity, the higher the SERS intensity for the Ag nanorod based SERS substrates. Accordingly, in order to further improve the SERS response of the Ag nanorod substrates, one can further increase the reflectivity of the underlayers through a proper surface coating such as multilayer dielectric coating39.

2.3. Enhancement of the sensitivity from periodical silver nanorods

Most studies have been focused on controlling the morphology of individual nanostructures to maximize their performance as SERS substrates, without consideration of the effect of the ordered arrangement of the nanostructure(s). Obviously, periodically arranged nanostructures might result in an enhancement in the localized electric field due to the resonance with the localized surface Plasmon [40,41] different from that of randomly arranged nanostructures, providing a possibility to improve the performance of SERS substrates. Therefore, it is of great interest to investigate the surface-enhanced Raman scattering from those nanostructures periodically arranged.

The substrates used in the experiment were N-type silicon substrates with an orientation of <100>. Hexagonal lattices (200 μm ×200 μm) of silicon patterns (~ 400nm in diameter) were fabricated on the substrates by electron beam lithography, where the separation distance of the patterns was controlled to be 0 (closely-packed), 50 nm, 100 nm, 200 nm, 300 nm and 400 nm, respectively. These were cleaned in sequence in acetone, ethanol, and de-ionized water baths supersonically, and were mounted on the substrates holder (cooled by liquid nitrogen) in a high vacuum e-beam deposition system (with a background vacuum level better than 2×10^{-5} Pa). Vertically aligned Ag nanorods were deposited on these substrates by the glancing-angle deposition (GLAD) technique described elsewhere [42]. During deposition, the substrate was cooled down to -20 °C, rotated at a speed of 2 rpm and its surface normal was set ~ 88° off the incoming vapor flux, and the deposition rate was monitored to be ~ 0.5 nm/s using a quartz crystal microbalance.

Oblique-view SEM images is shown from Fig.8 of Ag nanorods deposited on the same silicon substrate fabricated by electron beam lithography, with planar areas, and hexagonal lattice areas (~ 200 μm × 200 μm) of silicon patterns (~ 400nm in diameter) that are separated at 0 nm (closely packed), 50 nm, 100 nm, 200 nm and 300 nm, respectively. Silver nanorods grown on hexagonal lattice areas with other separations are of similar morphology and features. Since the height of silicon micro-patterns etched by electron beam lithography is only ~ 20 nm, there should be also deposition of Ag nanorods in the spaces among the patterns when their separation distance is large enough, e.g., 300 nm, see Fig.8(f). If the separation is not that large, there is normally no deposition of Ag nanorods in these areas because of the shadow effect by the GLAD technique [42]. It is found from the images that Ag nanorods were deposited onto the silicon patterns, forming two-dimensional hexagonal lattices of various lattice parameters (or separation distances of patterns). This provides us ideal samples to investigate the enhancement by the regular arrays to the Raman scattering.

We tested the performance of the hexagonal lattices of patterns covered by Ag nanorods as SERS substrates using R6G as the probing molecule. The samples are dipped in 1×10^{-6} M aqueous solution of R6G for 30 minutes and dried by a continuous gentle nitrogen stream. The Raman spectrum of R6G on each SERS substrate was obtained by measuring and averaging spectra from six different areas of the same hexagonal lattice, by a Reinshaw 100 Raman spectrometer using a 633 nm He-Ne laser as the excitation source. During measurements the spot size of the laser beam was defocused to ~ 10 μm in diameter, and the laser power was decreased to 0.47 mW to avoid any damage to the R6G molecules, signal accumunition time of 10 second per 600 cm^{-1}, 10x objective and NA 0.25. To make a clear comparison of the performance of these SERS substrates, the Raman spectrum of R6G from the unpatterned areas was also measured.

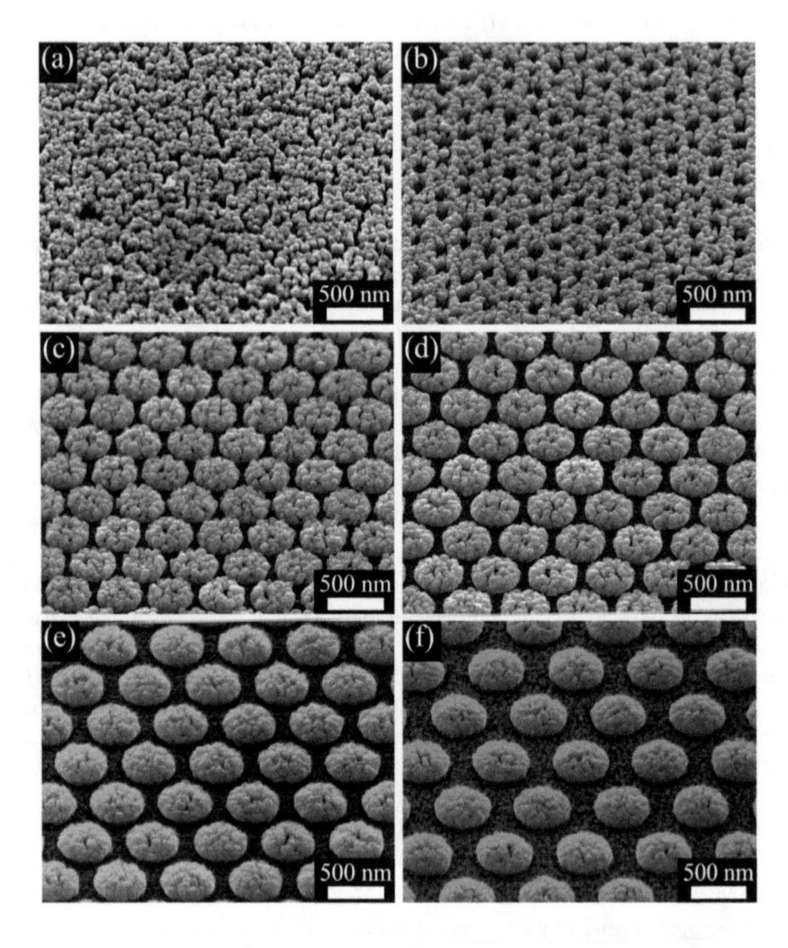

Figure 8. Olique-view SEM images of vertically aligned Ag nanorods deposited on (a) unpatterned silicon; and on hexagonal lattices of silicon patterns that are separated by (b) 0 nm (closely-packed); (c) 50 nm; (d) 100 nm; (e) 200 nm; and (f) 300 nm, respectively.

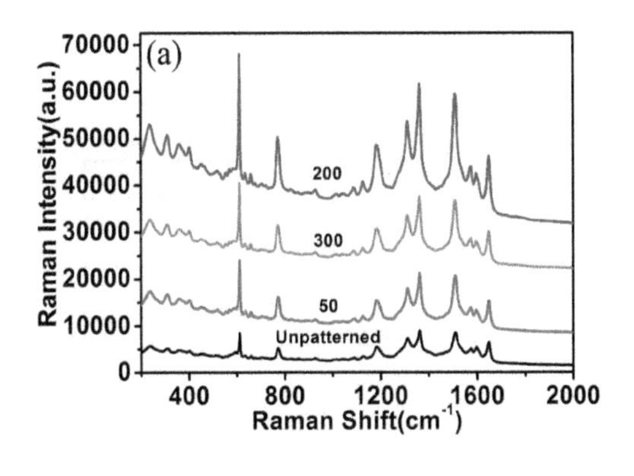

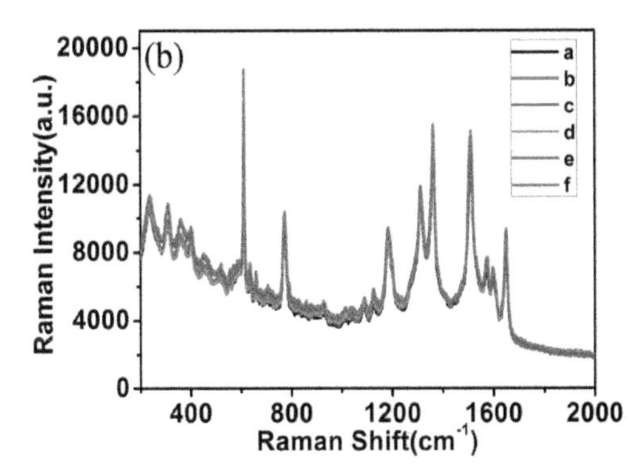

Figure 9. a) Raman spectra of R6G (concentration of 10^{-6} M) using aligned Ag nanorods deposited on unpatterned silicon substrates and on hexagonal lattices of silicon patterns separated by 50, 200 and 300 nm, respectively (To make each spectrum clear, the four spectrum is separated by adding value 0, 6500, 28000 and 14000, respectively), as SERS substrates; and (b) comparison of the Raman spectrum of R6G obtained from six different regions of the hexagonal lattice that are separated at 50 nm of aligned Ag nanorods.

Raman spectra is shown from Fig.9(a) of R6G molecules measured from different areas of the substrate, i.e., the unpatterned area (as the reference), and the hexagonal lattice areas of silicon patterns that are separated at 50 nm, 200 nm and 300nm. We see that each spectrum shows clearly the Raman features of the R6G molecules, and that the signal intensity of the Raman spectrum is very dependent on the substrate (or the separation distance of the patterns). For example, the signal intensity is the highest when the separation distance is 200 nm, and is the lowest in the unpatterned area. Difference in the Raman intensity between the unpatterned area and the hexagonal lattice areas is noticed. Fig.9(b) compares Raman spectra measured from six different areas of the closely-packed lattice, from which one sees that these spectra match quite well. This suggests a good homogeneity of the Raman signals from these SERS substrates.

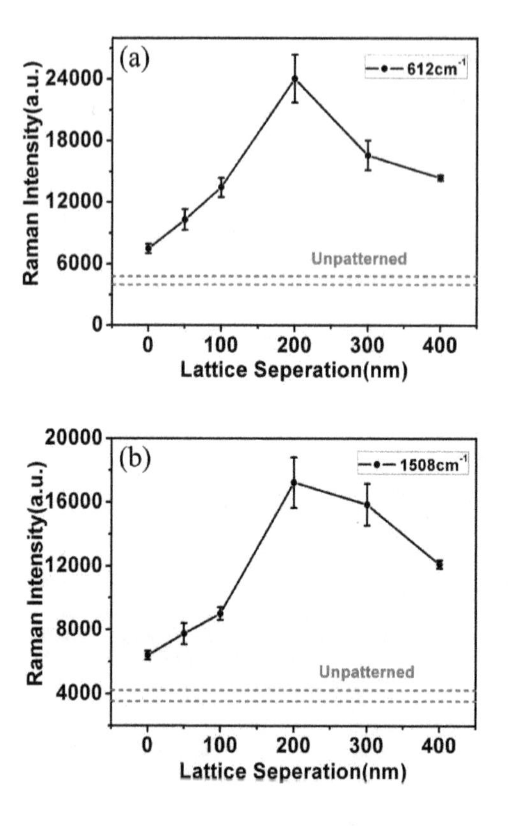

Figure 10. The intensity of the two characteristic peaks of R6G on hexagonal lattices of Ag nanorods: (a) 612 cm^{-1} and (b) 1508 cm^{-1}.

To gain a better understating of the latticing effect on the Raman scattering, we plot the intensity of the two characteristic peaks of R6G molecules, i.e. at ~ 612 cm^{-1} and ~ 1508 cm^{-1}, as a function of the separation distance of the lattice patterns, by comparing with the unpatterned Ag film (the two red dash line represent the sum and difference between the average and uncertainty of signal strength). Fig.10(a) and 10(b) shows respectively the intensity of the two peaks versus the separation distance of the lattice patterns. It is seen from the figures that for the two peaks, they demonstrated very similar dependence behavior on the separation distance. In comparison with signals from the unpartterned area, we notice that when the patterns are separated at ~ 200 nm, the signals are further enhanced by ~5 times compared with the unpatterned substrates. So the performance of silver nanostructure as SERS substrate is improved through periodical silver nanorods. [43]

3. Detection of PCBs at trace amount by SERS

With the highly sensitive SERS substrates described before, one can detect trace amount organic molecules by the SERS method.

SERS is extremely sensitive in water solutions, for water does not have any Raman peaks. When detecting organic pollutants in nonaqueous systems, we use volatile organic solvents, such as acetone, to dilute pollutants. As the organic solvent shows high Raman background, we need to make the solvent volatilizated completely before SERS measurement.

The powders of 2, 3, 3′, 4, 4′-pentachlorinated biphenyl used in this study were commercially available fromthe AccuStandard Company. Since there is no Raman data of 2, 3, 3′, 4, 4′-pentachlorinated biphenyl reported, we first measured its Raman spectrum and that of acetone, for comparison, see Fig 11(a). To clearly show most characteristic peaks of 2, 3, 3′, 4, 4′-pentachlorinated biphenyl, the Raman spectrum was plotted in two regions of 300 to 1000 cm^{-1} and 1000 to 1700 cm^{-1} respectively, see Figs 11(b) and 14(c). From the Figs one sees that the strongest peaks are located at 342, 395, 436, 465, 495, 507, 517, 598, 679, 731, 833, 891, 1032, 1136, 1179, 1254, 1294, 1573, and 1591 cm^{-1}, respectively; while for acetone the characteristic peaks are at 530, 786, 1065, 1220, 1428 and 1709 cm^{-1}, respectively. It is suggested that 2, 3, 3′, 4, 4′-pentachlorinated biphenyl is distinguishable from acetone and that acetone can be used as the solvent for the SERS measurements, as 2, 3, 3′, 4, 4′-pentachlorinated biphenyl is not soluble in water.

Because the SERS sensitivity is also dependent on the sample treatment, we employed in this study a very simple method to prepare SERS samples, i.e. dropping a small volume (~ 0.5 uL) of solutions of 2, 3, 3′, 4, 4′-pentachlorinated biphenyl in acetone on Ag nanorods using a top single channel pipettor and then blowing away the acetone with a continuous, gentle nitrogen blow. Fig 12(a) shows the Raman spectra of 2, 3, 3′, 4, 4′-pentachlorinated biphenyl dissolved in acetone at concentrations of 10^{-4} to 10^{-10}mol/L, respectively. The accumulation time of each Raman spectrum was 50 seconds and we used only 1% laser power to avoid changing of pentachlorinated biphenyl. When the small volume (~ 0.5 uL) of solutions of 2, 3, 3′, 4, 4′-pentachlorinated biphenyl was dropped on Ag nanorods, it became a circular spot with diameter of about 4 mm. The Raman spectrum was accumulated from a 2 um diameter circular area on the substrates. Therefore, for the solution at concentration of 10^{-10}mol/L, only about ten 2, 3, 3′, 4, 4′-pentachlorinated biphenyl molecules ($2*10^{-23}$mol) would be accumulated in SERS; if that was at concentration of 10^{-8}mol/L, about 1000 molecules ($2*10^{-21}$mol) would be accumulated and so on.

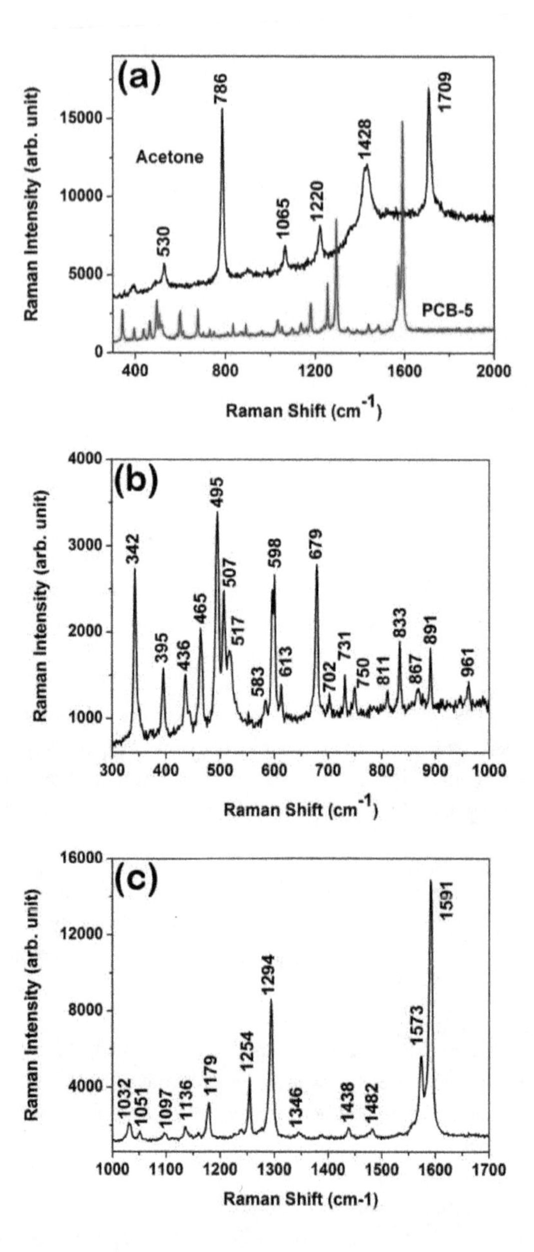

Figure 11. a) Comparison of Raman spectra of PCB-5 powders and acetone; (b) and (c) show details of the Raman spectrum of PCB-5 powders.

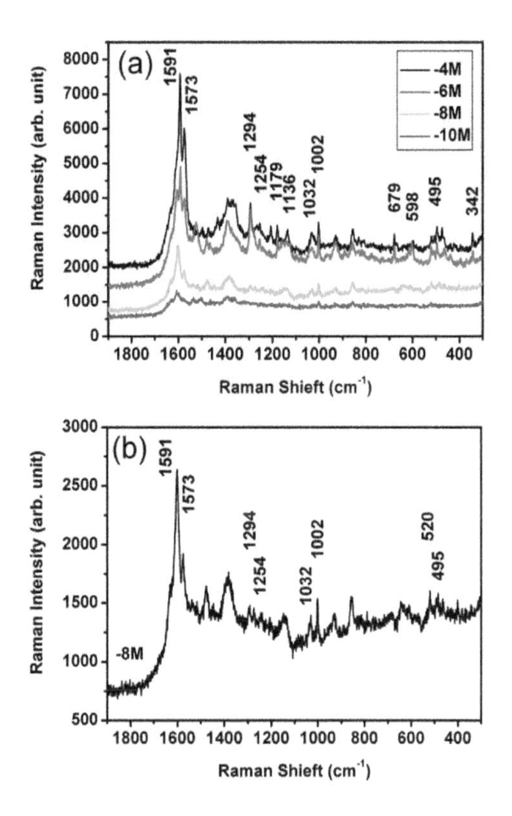

Figure 12. a) SERS spectra of PCB-5 dissolved in acetone with various concentrations; (b) SERS spectrum of PCB-5 in acetone at a concentration of 10^{-8}mol/L.

One sees that the Raman peaks of 10^{-4} mol/L PCB-5 solution located at 342, 495, 598, 679, 1032, 1136, 1179, 1254, 1294, 1573 and 1591 cm^{-1} march the Raman peaks of powder PCB-5 very well, this is quite different from the characteristic peaks of acetone. The peak around 1390 cm^{-1} represents disordered and amorphous carbon on the substrates. Fig 12(b) shows the SERS spectra of 10^{-8} mol/L PCB-5. Peaks located at 495, 1032, 1294, 1573 and 1591 cm^{-1} can march the Raman peaks of powder PCB-5. It indicates that the peaks shown in Fig 12(a) and (b) are the characteristic peaks of dissolved PCB-5, and PCB-5 with a concentration of 10^{-8}mol/L can be detected by the SERS method in the authors' work.

Large scale arrays of aligned and well separated single crystalline Ag nanorods on planar silicon substrate can be fabricated by GLAD method and these Ag films can be used as SERS substrates. With these substrates 2, 3, 3′, 4, 4′- PCB-5 molecules were detected even at a concentration of 10^{-8}mol/L by the SERS method, which indicates that trace amount of PCBs can be detected by the SERS method with Ag nanorods as SERS substrates [44].

4. Detection of POPs at trace level in real environmental samples

In sections 3 we introduced the SERS method to detect trace amount PCBs. In those experiments, the PCBs are in acetone solutions, as fundamental study. In this section we introduce some examples in practical trace POPs detection.

4.1. Detection of PCBs in dry soil samples

The polluted soil samples were dried and made into small powers which were acquired from the Nanjing Institute of Soil (China). With a combination of the high-resolution gas chromatography and mass spectrometry techniques, sample I proved to contain about 5 μg/g PCBs and sample II proved to contain about 300 μg/g PCBs. 0.2 g soil sample I was put into 20 mL acetone and was agitated uniformly for about 5 minutes. This suspension was precipitated for 30 minutes and the transparent acetone solution in the upper layer was taken as solution sample A. 0.2 g soil sample I was put into 200 mL acetone and solution sample B was obtained through the aforementioned process. 0.2 g soil sample II was put into 20 mL acetone to obtain solution sample C and was put into 200 mL acetone to obtain solution sample D.

The Ag nanorods SERS substrates were put into the solution samples A, B, C and D, respectively. After 30 minutes, the Ag nanorods substrates were taken out of the solutions and the acetone on the substrates was blown away using a nitrogen flow. The Raman spectra of these substrates dipped into solution samples were measured by a Renishaw Raman 100 spectrometer using a 633 nm He-Ne laser as the excitation source at room temperature.

Figs 13(a), (b), (c) and (d) show the measured Raman spectrum of the Ag substrates dipped into sample A, B, C and D, respectively. From Figs 13 (a), (b) and (c), one sees peaks at ~1600, 1280, 1240, 1150, 1030 and 1000 cm^{-1} clearly, demonstrating the common feature of PCBs. The peaks around 1590~1600cm^{-1} present benzene stretching vibration mode; the peak around 1280cm^{-1} presents CC bridge stretching vibration mode; the peak around 1030cm^{-1} presents CH bending in-plane mode; the peak around 1000cm^{-1} presents trigonal breathing vibration mode; and peaks around 1240~1250cm^{-1} and 1140~1200cm^{-1} present the vibration peaks induced by Cl substituent. These characteristic peaks suggest that PCBs in dry soil can be detected by the SERS method by dissolving into acetone. The most widely used PCBs are trichlorobiphenyls and pentachlorobiphenyls, we assumed that the molecular weight of the PCBs in the soil samples is 300, then the concentration of the PCBs acetone solution in solution sample A, B, C and D are about 10^{-5}mol/L, 10^{-5}mol/L, 10^{-6}mol/L, 10^{-7}mol/L, and 10^{-8}mol/L, respectively. Thus, with silver nanorod substrates, 5ug/g PCBs in dry soil samples can be detected by the SERS method.

4.2. Detection of PCBs in white spirit

PCBs in white spirit can also be detected by the SERS method with silver nanorod substrates. The concentration of PCBs in white spirit is about 10^{-4}mol/L. We put a drop of PCBs "polluted" white spirit on the silver nanorod substrates and made the white spirit volatilized away. Then, we found PCBs Raman signal with the SERS method described before. Figs 14 (a) and (b) show the SERS spectra of pure white spirit and white spirit with 10^{-4}mol/L PCBs, respectively. One can recognize characteristic Raman peaks of PCBs around 1590, 1290, 1240, 1030 and 1000 cm^{-1} in Fig 14 (b).

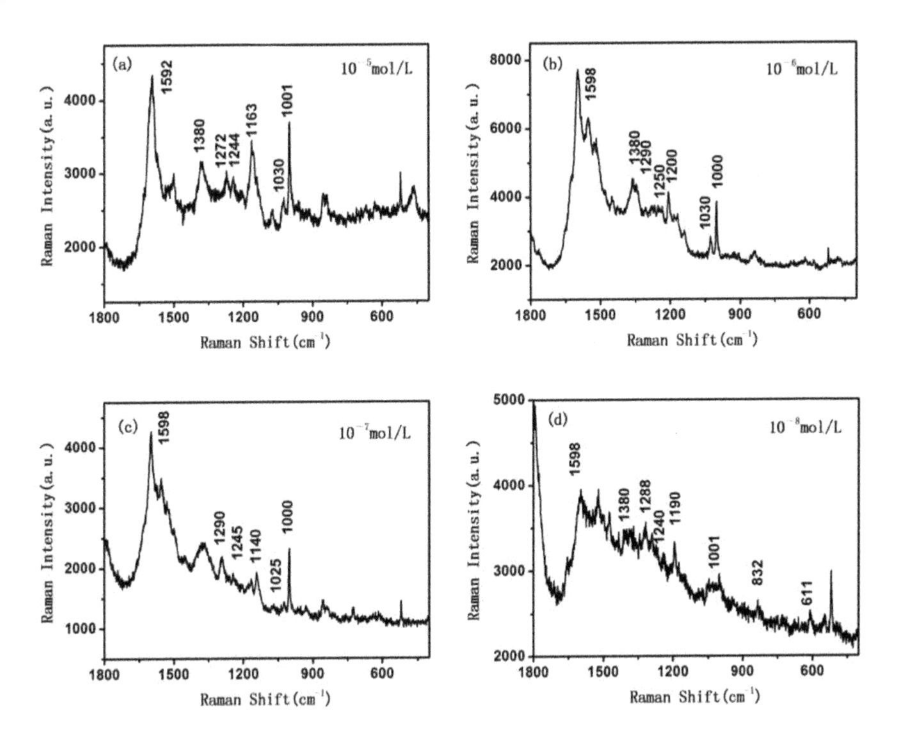

Figure 13. SERS spectra of PCBs in dry soil samples after being treated by acetone: (a) ~10^{-5}mol/L; (b) ~10^{-6}mol/L; (c) ~10^{-7}mol/L; (d) ~10^{-8}mol/L.

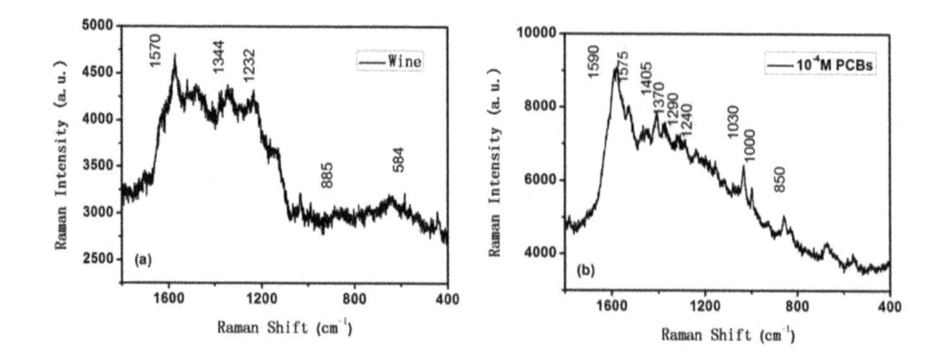

Figure 14. SERS spectra of white spirit without and with PCBs.

4.3. Detection of Melamine in milk

In the year 2009, milk produced by Sanlu Co. (China) was found to contain amounts of Melamine in much higher concentrations than usual. Milk with Melamine seems to contain more protein when detecting nitrogen concentration, but it is poisonous to children. With the SERS method with silver nanorods as substrates, we detected Melamine in milk. Figs 15 (a) and (b) show the SERS spectra of pure Melamine and milk with Melamine.

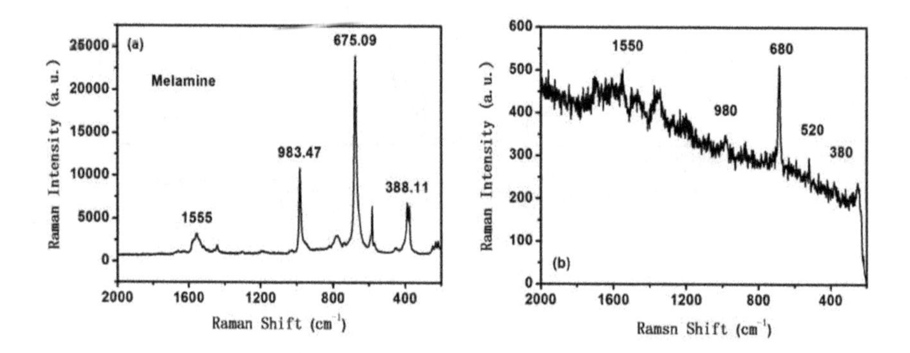

Figure 15. SERS spectra of Melamine and trace Melamine in milk.

5. Summary

In this chapter, it is introduced that although persistent organic pollutants such as PCBs are difficult to detect at trace amount, they can be detected and recognized rapidly via the SERS technique. Ag nanostructured SERS substrates prepared by the glancing angle deposition method are excellent at detection and their sensitivity can be further improved by tuning the thin underlayer films. With well designed and prepared Ag nanostructured SERS substrates, pentachlorinated biphenyl molecules are detected and recognized at trace level using the SERS method. These series of studies provide a potential method for trace pollutant detection via silver nanostructure.

Author details

Zhengjun Zhang[1], Qin Zhou[1,2] and Xian Zhang[1]

*Address all correspondence to: zjzhang@mail.tsinghua.edu.cn

1 Advanced Materials Laboratory, Department of Materials Science and Engineering, Tsinghua University, Beijing, P. R. China

2 Institute of nuclear and new energy technology, Tsinghua University, Beijing, P. R. China

References

[1] Ross G. The public health implications of polychlorinated biphenyls (pcbs) in the environment.Ecotoxicology and Environmental Safety, 2004, 59(3): 275–291.

[2] Ohtsubo Y, Kudo T, Tsuda M, et al. Strategies for bioremediation of polychlorinated biphenyls. Applied Microbiology and Biotechnology, 2004, 65(3): 250–258.

[3] Cicchetti D V, Kaufman A S, Sparrow S S. The relationship between prenatal and postnatal exposure to polychlorinated biphenyls (pcbs) and cognitive, neuropsychological, and behavioral deficits: a critical appraisal. Psychology in the Schools, 2004, 41(6): 589–624.

[4] Hong J E, Pyo H, Park S J, et al. Determination of hydroxy-pcbs in urine by gas chromatography/mass spectrometry with solid-phase extraction and derivatization. AnalyticaChimicaActa, 2005, 531(2): 249–256.

[5] Namiesnik J, Zygmunt B. Selected concentration techniques for gas chromatographic analysis of environmental samples. Chromatographia, 2002, 56Suppl. S: S9–S18.

[6] Pitarch E, Serrano R, Lopez F J, et al. Rapid multiresidue determination of organochlorine and organophosphorus compounds in human serum by solid-phase extrac-

tion and gas chromatography coupled to tandem mass spectrometry. Analytical and Bioanalytical Chemistry, 2003, 376(2): 189–197.

[7] Barra R, Cisternas M, Suarez C, et al. Pcbs and hchs in a salt-marsh sediment record from south-central Chile: use of tsunami signatures and cs-137 fallout as temporal markers. Chemosphere, 2004, 55(7): 965–972.

[8] Moskovits M. Surface-enhanced spectroscopy. Reviews of Modern Physics, 1985, 57(3): 783–826.

[9] Zhou Q, Li Z C, Yang Y, et al. Arrays of aligned, single crystalline silver nanorods for trace amount detection. Journal of Physics D-Applied Physics, 2008, 41(15200715).

[10] Kudelski A. Analytical applications of raman spectroscopy. Talanta, 2008, 76(1): 1–8.

[11] Zhang X Y, Zhao J, Whitney A V, et al. Ultrastable substrates for surface-enhanced raman spectroscopy: al2o3 overlayers fabricated by atomic layer deposition yield improved anthrax biomarker detection. Journal of the American Chemical Society, 2006, 128(JA063876031): 10304–10309.

[12] Tan R Z, Agarwal A, Balasubramanian N, et al. 3d arrays of sers substrate for ultrasensitive molecular detection. Sensors and Actuators a-Physical, 2007, 139(1-2Sp. Iss. SI): 36–41.

[13] Isola N R, Stokes D L, Vo-Dinh T. Surface enhanced raman gene probe for hiv detection. Analytical Chemistry, 1998, 70(7): 1352–1356.

[14] Vodinh T, Houck K, Stokes D L. Surface-enhanced raman gene probes. Analytical Chemistry, 1994, 66(20): 3379–3383.

[15] Tripp R A, Dluhy R A, Zhao Y P. Novel nanostructures for sersbiosensing. Nano Today, 2008, 3(3-4): 31–37.

[16] Chaney S B, Shanmukh S, Dluhy R A, et al. Aligned silver nanorod arrays produce high sensitivity surface-enhanced raman spectroscopy substrates. Applied Physics Letters, 2005, 87(0319083).

[17] Zhang Z Y, Zhao Y P. Tuning the optical absorption properties of agnanorods by their topologic shapes: a discrete dipole approximation calculation. Applied Physics Letters, 2006, 89(0231102).

[18] Malac M, Egerton R F, Brett M J, et al. Fabrication of submicrometer regular arrays of pillars and helices. Journal of Vacuum Science & Technology B, 1999, 17(6): 2671–2674.

[19] Dick B, Brett M J, Smy T. Investigation of substrate rotation at glancing-incidence on thin-film morphology. Journal of Vacuum Science & Technology B, 2003, 21(6): 2569–2575.

[20] Dick B, Brett M J, Smy T J, et al. Periodic magnetic microstructures by glancing angle deposition. Journal of Vacuum Science & Technology A-Vacuum Surface and Films, 2000, 18(4Part 2): 1838–1844.

[21] Alouach H, Fujiwara H, Mankey G J. Magnetocrystalline anisotropy in glancing angle deposited permalloy nanowire arrays. Journal of Vacuum Science & Technology A, 2005, 23(4): 1046–1050.

[22] Singh J P, Tang F, Karabacak T, et al. Enhanced cold field emission from 100 oriented beta-w nanoemitters. Journal of Vacuum Science & Technology B, 2004, 22(3): 1048–1051.

[23] Hawkeye M M, Brett M J. Glancing angle deposition: fabrication, properties, and applications of micro- and nanostructured thin films. Journal of Vacuum Science & Technology A, 2007, 25(5): 1317–1335.

[24] Leverette C L, Shubert V A, Wade T L, et al. Development of a novel dual-layer thick ag substrate for surface-enhanced raman scattering (sers) of self-assembled monolayers. Journal of Physical Chemistry B, 2002, 106(34): 8747–8755.

[25] Li H G, Cullum B M. Dual layer and multilayer enhancements from silver film over nanostructured surface-enhanced raman substrates. Applied Spectroscopy, 2005, 59(4): 410–417.

[26] Li H G, Baum C E, Sun J, et al. Multilayer enhanced gold film over nanostructure surface-enhanced raman substrates. Applied Spectroscopy, 2006, 60(12): 1377–1385.

[27] Yang Y A, Bittner A M, Kern K. A new sers-active sandwich structure. Journal of Solid State Electrochemistry, 2007, 11(2): 150–154.

[28] Mulvaney S P, He L, Natan M J, et al. Three-layer substrates for surface-enhanced raman scattering: preparation and preliminary evaluation. Journal of Raman Spectroscopy, 2003, 34(2): 163–171.

[29] Driskell J D, Shanmukh S, Liu Y, et al. The use of aligned silver nanorod arrays prepared by oblique angle deposition as surface enhanced raman scattering substrates. Journal of Physical Chemistry C, 2008, 112(4): 895–901.

[30] Driskell J D, Lipert R J, Porter M D. Labeled gold nanoparticles immobilized at smooth metallic substrates: systematic investigation of surface plasmon resonance and surface-enhanced raman scattering. Journal of Physical Chemistry B, 2006, 110(35): 17444–17451.

[31] Addison C J, Brolo A G. Nanoparticle-containing structures as a substrate for surface-enhanced raman scattering. Langmuir, 2006, 22(21): 8696–8702.

[32] Misra A K, Sharma S K, Kamemoto L, et al. Novel micro-cavity substrates for improving the raman signal from submicrometer size materials. Applied Spectroscopy, 2009, 63(3): 373–377.

[33] Shoute L, Bergren A J, Mahmoud A M, et al. Optical interference effects in the design of substrates for surface-enhanced raman spectroscopy. Applied Spectroscopy, 2009, 63(2): 133–140.

[34] Liu Y J, Zhao Y P. Simple model for surface-enhanced raman scattering from tilted silver nanorod array substrates. Physical Review B, 2008, 78(0754367).

[35] Mitsas C L, Siapkas D I. Generalized matrix-method for analysis of coherent and incoherent reflectance and transmittance of multilayer structures with rough surfaces, interfaces, and finite substrates. Applied Optics, 1995, 34(10): 1678–1683.

[36] Fu J X, Park B, Zhao Y P. Nanorod-mediated surface plasmon resonance sensor based on effective medium theory. Applied Optics, 2009, 48(23): 4637–4649.

[37] Abell J L, Driskell J D, Dluhy R A, et al. Fabrication and characterization of a multiwell array sers chip with biological applications. Biosensors & Bioelectronics, 2009, 24(12): 3663–3670.

[38] Yang W H, Hulteen J, Schatz G C, et al. A surface-enhanced hyper-raman and surface-enhanced raman scattering study of trans-1,2-bis(4-pyridyl)ethylene adsorbed onto silver film over nanosphere electrodes. Vibrational assignments: experiment and theory. Journal of Chemical Physics, 1996, 104(11): 4313–4323.

[39] Zhou Q, Liu Y J, He Y P, et al. The effect of underlayer thin films on the surface-enhanced raman scattering response of agnanorod substrates. Applied Physics Letters, 2010, 97(12190212).

[40] Brown R. and Milton M., Nanostructures and nanostructured substrates for surface —enhanced Raman scattering. J. Raman Spectrosc. 2008. 39, 1313.

[41] HaynesC L, McFarland A D and Van Duyne R P,Surface-enhanced Raman spectroscopy, Anal. Chem. 2005, 77, 338A.

[42] Zhou Q., Li Z .C., Yang Y. and Zhang Z.J., Arrays of aligned single crystalline silver nanorods for trace amount detection, J. Phys. D Appl. Phys. 2008.41(152007).

[43] Zhang X, Zhou Q, Wang W P, Shen L, et al.Latticing vertically aligned Ag nanorods to enhance its SERS sensitivity.Materials Research Bulletin, 2012, 47,921~924.

[44] Zhou Q, Yang Y, Ni J, et al. Rapid detection of 2, 3, 3 ', 4, 4 '-pentachlorinated biphenyls by silver nanorods-enhanced raman spectroscopy. Physica E-Low-Dimensional Systems & Nanostructures, 2010, 42(5): 1717–1720.

[45] Michaels A M, Jiang J, Brus L. Ag nanocrystal junctions as the site for surface-enhanced raman scattering of single rhodamine 6g molecules. Journal of Physical Chemistry B, 2000, 104(50): 11965–11971.

[46] Mcfarland A D, Young M A, Dieringer J A, et al. Wavelength-scanned surface-enhanced raman excitation spectroscopy. Journal of Physical Chemistry B, 2005, 109(22): 11279–11285.

[47] Qin L D, Zou S L, Xue C, et al. Designing, fabricating, and imaging raman hot spots. Proceedings of the National Academy of Sciences of the United States of America, 2006, 103(36): 13300–13303.

[48] Liu Y J, Zhang Z Y, Zhao Q, et al. Surface enhanced raman scattering from an agnanorod array substrate: the site dependent enhancement and layer absorbance effect. Journal of Physical Chemistry C, 2009, 113(22): 9664–9669.

[49] Gansel J K, Thiel M, Rill M S, et al. Gold helix photonic metamaterial as broadband circular polarizer. Science, 2009, 325(5947): 1513–1515.

Chapter 2

The Comparison of Soil Load by POPs in Two Major Imission Regions of the Czech Republic

Radim Vácha, Jan Skála, Jarmila Čechmánková and
Viera Horváthová

Additional information is available at the end of the chapter

1. Introduction

The Czech Republic belongs to the countries with long-term industrial history. The environmental load by persistent organic pollutants pollution has been proved to follow the industrial development, especially concerning the polycyclic aromatic hydrocarbons emissions. In the Czech Republic the industrial growth started during 19th century and in the beginning of 20th century at the time of the Austro-Hungarian Monarchy. The industrial development continued after the Monarchy collapse and the Czechoslovak Republic formation (1918 - 1938). The rapid industry growth was led by heavy industry priority in the period of socialistic economy (1948 – 1990) and caused the wide environmental damages. The imission outputs reached maximum in 70th years when the daily average concentrations of SO_2 (gaseous emission) were over $50 \mu g/m^3$ and in the coalfield areas of North Bohemia up to 70 - $100 \mu g/m^3$ following the data of Czech Hydrometeorological Institute [1]. The loading by floating dust particles was more than 70 - $100 \mu g/m^3$ and in extreme cases reached $150 \mu g/m^3$.

There are two main coal mining regions in the Czech Republic (see Figure 1.). The history of brown coal mining started in North Bohemian Region in the beginning of 19th century (1819) and reached the maximum in the eighties of 20th century. The history of black coal mining in North Moravian Region is very similar and the mining activity peaked in the eighties of 20th century (about 20 millions tons per year). Opencast coal-mining activity in North Bohemia has been changing the landscape character in more intense way, however the deep mines in North Moravia has also caused environmental damages due to terrain subsidence and lagoons with coal powder and waste. Other important risk are linked to the combustion of brown coal of low quality with increased contents of sulphur and arsenic [2] in coal-fired power stations in North Bohemia and to the presence of metallurgical industry in the North

Moravian region. The load of both areas by risky elements and persistent organic pollutants gave them the designation of imission regions. The North Bohemian region covers the area of 5 districts (Decin, Teplice, Usti nad Labem, Most and Chomutov and neighbouring districts in the West Bohemian region where increased load still remains). The region is spread along the Czech-German border shaped by the Ore Mountains. The North Bohemian basin is delimitated by the dislocation at the foothill of the Ore Mountains. The North Moravian region situated close to Czech-Polish border covers the area of 3 districts (Ostrava, Karvina, Frydek-Mistek). The flat character of the landscape in west part of the region (Karvina, Ostrava) passes to mountainous area forming the Czech-Slovak borderland (the Moravsko-slezske Beskydy Mountains). The load of environment in both regions is historically increased with the historical pollution peak in seventies and eighties of 20th century when high content of emission-out puts in the air connected with acid rains led to perceptible damages of the environment (especially damage of the spruce forest in the Ore Mountains). The situation started to change after 1990 thanks to industrial production decrease and the necessity of technology improvement of coal-fired power stations (the installation of efficient dust particles filters in the beginning of the 21st century). The modernization of four coal-fired power stations situated in the North Bohemian region (Ledvice, Pocerady, Tusimice and Prunerov) has been approaching in two periods. In the 1st period (1996 – 1999) there were radically decreased the emission out puts in following extent: SOx -92%, NOx -50%, CO -77% and solid polluting particles -93%. The next period of modernization is running and will be finished till 2020 following precise schedule of the works. The next decrease of emission out puts will be reached in the following extent: SOx -57%, NOx -59%, CO_2 -31% and solid polluting particles -39%, data from [3].

The comfortless situation remains in the North Moravian region and increased contents of emission out-puts in the air are the theme of many professional and public discussions [4,5]. The special attention is paid to increased content of polycyclic aromatic hydrocarbons in environment which is connected with increased number of some inhabitant's diseases in the region [3]. The load stemmes from emission out puts from heavy industrial factories (for example the Trinec and Ostrava ironworks factories etc). There is one coal-fired power station (Detmarovice) in the region where black coal is used. The mountainous area situated in eastern part of the region serve for recreation and sport and there are no important sources of pollution. Nevertheless the increased pollution was proved in the mountains are due to imission out puts from western part of the region. Other environmental hazards in the region are linked to the existence of lagoons where around 300,000 tons of petroleum sludge have been deposited. The recent dredge and liquidation of sludge meets some technical and economical inconveniences.

The soil is one of the important environmental sinks of pollution and soil contamination can reflect long-term load by dry and wet depositions. Increased soil load by risky substances poses serious threats to environment, plant production and food security. The maintenance of suitable state of soil load by risky substances should be an interest of every society. The evaluation of soil load by risky substances must be supported by the knowledge of risky substances background values, their inputs into soils, their behaviour and fate in the soil environment, their transfer into the plants etc. The approaches to limit values system are not unified across the world, nor in European context and different philosophies may be used for the evaluation of soil contamination. There has been paid longterm attention to soil contamination issue in the Czech Republic. The potentially toxic compounds observed in Czech agricultural soils can be separated into two main groups of pollutants:

- Inorganic pollutants - potentially risky elements (REs), As, Be, Cd, Co, Cr, Hg, Cu, Mn, Ni, Pb, V, Zn, respectively Se and Tl
- Organic pollutants – persistent organic pollutants (POPs), A wide group of different organic substances, with linear or cyclic character. The current list of POPs observed in Czech legislation (Soil Protection Act) includes monocyclic and polycyclic hydrocarbons, PCBs, sum of DDT and petroleum hydrocarbons (table 1).

POPs
Monocyclic aromatic hydrocarbons
benzene, toluene, xylene, ethylbenzene
Polycyclic aromatic hydrocarbons
naphtalene, anthracene, pyrene, phluoranthene, phenanthrene, chrysen, benzo(b)phluoranthene, benzo(k)phluoranthene, benzo(a)anthracene, benzo(a)pyrene, indeno(c,d)pyrene, benzo(ghi)perylene
chlorinated hydrocarbons
PCB(28+52+101+118 +138+153+180), HCB, α-HCH, β-HCH, γ-HCH
Pesticides
DDT, DDD, DDE
Others
styrene, petroleum hydrocarbons
PCDF
2,3,7,8 TeCDF, 1,2,3,7,8 PeCDF, 2,3,4,7,8 PeCDF, 1,2,3,4,7,8 H_xCDF, 1,2,3,6,7,8 H_xCDF, 1,2,3,7,8,9 H_xCDF, 2,3,4,6,7,8 H_xCDF, 1,2,3,4,6,7,8 H_pCDF, 1,2,3,4,7,8,9 H_pCDF, OCDF PCB 189, PCB 170, PCB 180
PCDD
2,3,7,8 TeCDD, 1,2,3,7,8 PeCDD, 1,2,3,4,7,8 H_xCDD, 1,2,3,6,7,8 H_xCDD, 1,2,3,7,8,9 H_xCDD, 1,2,3,4,6,7,8 H_pCDD,OCDD

Table 1. Persistent organic pollutants observed in Czech agricultural soils

The system of limit values of soil contamination must accept sources of risky substances that influence the behaviour of risky substances in the soil (mobility, bioavailability). POPs can originate from:

- Natural sources - volcanic activity (REs, POPs), natural fires (POPs) etc.
- Anthropogenic sources – like industrial activities, transport emissions, the use of agrochemicals and biosolids in agriculture, waste water production etc.

Increased inputs of potentially toxic compounds into the soils may result in soil contamination that may negatively influence:

- The ecosystem - soil functions, contamination of aquatic systems, plants, animals etc.
- Plant production – the quantity and quality.
- Human health – via contamination of food chain, dermal or inhalation intake etc.

The efficient ways of the control and regulation of risky substances in the soil are legislatively mandatory limit values. The limit values of risky substances in the soil are derived on the basis of:

- Real state of soil load by risky substances reflected natural and anthropogenic diffuse load. Limit values of this kind are usually specified as "background values" of risky substances in the soil, [6] and [7].

- Experimentally derived values, that are focused on target risk following from soil use and observed environmental component (the damage of quantity and quality of plant production, the reduction of soil microbial activity etc.).

One of the most effective and sophisticated limit values systems is so called hierarchical limit values system that should be able to register target risk following the soil contamination. This system is usually used in many European countries (Germany, Netherlands and Switzerland) as system of "A, B and C limits" where

A – represents background values of risky substances in the soil. Generally, this limit value fulfils the principal of precaution.

B – is focused on target risk. The limit may be targeted on the quality or quantity of plant production (this approach is used rarely and is determined rather for small allotment producers than for agriculture) or on the decreasing of soil microbial activity and soil transformation functions etc.

C – is used as remediation (decontamination) limit that is based on the human health risk or environmental damage.

The limit values system focused on remediation needs (in the order of C limit level) are used in some countries (Great Britain, USA – EPA methodology, [8]). Given limit values of risky substances delimit risky substances concentrations that may distinctly affect human health or environment. After the exceeding of this limit the site-based risk assessment must be done and the results of risk assessment study determine next approach (the remediation, land use change). The proposal of the EU Soil Protection Act [9] is based on similar philosophy. Three steps are required on national level of member countries:

- The elaboration of soil contaminated sites register.

- The realisation of risk assessment studies on contaminated sites.

- The realisation of remediation approaches.

The directive of the Ministry of Environment of the Czech Republic No. 13/1994 Coll. [10] regulates the contents of REs and POPs in Czech agricultural soils. The limits of REs are determined for light texture soils and the other soils in the form of the aqua regia extract and the extract in 2M HNO_3 (cold method). The limit values for POPs are determined for the groups of monocyclic aromatic hydrocarbons, polycyclic aromatic hydrocarbons, chlorinated hydrocarbons (including pesticides) and petroleum hydrocarbons. All the limit values are defined as tolerable contents of risky substances but there is no relationship to any actual risk. It brings difficulties by the evaluation and interpretation of soil load by risky substances in many cases. Moreover the limits for POPs were assessed on the base of the values taken from Nederland because no actual values of Czech soil load by POPs were available in 1994. Two years later were published the first real data about soil load by POPs in the Czech Republic [5]. Some limit values in the directive No. 13/1994 Coll. are too low (especially some individual PAHs) because they stemm from Dutch legislation limits derived for sandy soils with low content of soil organic matter and are not suitable for Czech soils of different properties. There naturally arises the task of the proposal of the directive No.13 amendment based on the principal of hierarchical limit values system [11].

Three levels of the limits has been proposed:

Prevention limit – based on background values of risky substances in Czech agricultural soils. Prevention limit were proposed for REs and POPs. The exceeding of the limit shows increased anthropogenic soil load by risky substances. From the viewpoint of limit interpretation it is prohibited to use the sludge or sediment for soil fertilization in the case of limit exceeding. The proposed prevention limits for POPs based on the actual load of Czech agricultural soils [5] are presented in table 2.

Indication limit – was derived experimentally and the limit exceeding indicates the risk of increased REs transfer from the soil into the plants. Indication limit was proposed for REs only. The more detailed assessment is recommended at the locality after exceeding of indication limit. In the case of POPs the transfer from soil into plants via root intake is limited. More individual risks must be accepted for POPs indication limit and the realisation of risk assessment is recommended on the field seriously contaminated by POPs (in term of C limit level). In spite of these facts the simplified indication limits for some POPs were proposed [12], table 3. The limit values were determined as the lowest contents of risky substances in the soil that may cause any health risk. The transfer of risky substance from the soil to human bodies by dermal, oral and dietary intake was accepted.

Decontamination limit – has not been proposed yet.

The proposal of the directive No. 13/1994 Coll. amendment brings new approach to soil load by risky substances evaluation and its commencement could improve the management of contaminated sites. It could be very useful tool for the soil management in the imission regions of North Bohemia and North Moravia.

This report compares the load of two imission regions of the Czech Republic by three types of persistent organic pollutants: polycyclic aromatic hydrocarbons (PAHs), polychlorinated biphenyls (PCBs) and DDTs (DDT, DDE and DDD). Although PCBs and DDTs have not been used in Europe since eighties, the load of soil by both groups of pollutants is still increased.

2. Materials and methods

2.1. The characterization of study area

The area of the North Bohemian region and North Moravian region is presented in Figure 1. The area of North Bohemian region is 3,184.65 km^2 and the area of North Moravian Region is 1,834.25km^2.

2.1.1. North Bohemian region

2.1.1.1. Districts

The region comprises 5 districts (NUTS 5 level): Decin, Usti nad Labem, Teplice, Most and Chomutov.

District Decin is situated on the north of region and its area is 908.58km^2. In the district live 132,718 inhabitants in 52 municipalities, 14 of which are classified as towns. The capital of district is the town Decin.

The density of population is 146 inhabitants/km^2 in the district. The agricultural land covers 40.1% of the district area and the ratio of arable land is 32.94%, it means 13.21% of the district area. The other land covers 59.9% of district area and the ratio of forest is 82.25%, this is 49.27% of district area. There are 4 geomorphologic formations in district area, Decin Upland (north-western and central part), Central Bohemian Uplands (south and south-western part), Luzicke Mountains (eastern part) and Sluknov Downs (northern part). The highest point (Penkavci vrch) has the altitude of 792 m.a.s.l. and the lowest point (the Elbe river bank in Hrensko on the border with Germany) that is the lowest point of the Czech Republic also has the altitude of 115 m.a.s.l. The Elbe river (the biggest Czech river) traverses in the west part of the district.

District Usti nad Labem is situated in south-west direction from the Decin district and its area is 404,45km^2. In the district live 118,194 inhabitants and 84.44% of them live in towns. The capital of district is Usti nad Labem. The density of population has value of 292 inhabitants/km^2 in the district. The agricultural land covers 45.66% of district area and the ratio of arable land is 29.33%, it means 13.39% of district area. The other land covers 54.34% of district area and the ratio of forest is 57.72%, this is 31.37% of district area. There are 2 most important geomorphological formations, Central Bohemian Uplands (south-western and western part) and the Ore Mountains (northern part along the border with Germany). The Elbe river flows through the district.

POPs	Preventive value (µg/kg of dry matter)
Monocyclic aromatic hydrocarbons	
Benzene	30
Toluene	30
Xylene	30
Styrene	50
Ethylbenzene	40
Polycyclic aromatic hydrocarbons	
Fluoranthene	300
Pyrene	200
Phenanthrene	150
Benzo(b)fluoranthene	100
Benzo(a)anthracene	100
Anthracene	50
Indeno(cd)pyrene	100
Benzo(a)pyrene	100
Benzo(k)fluoranthene	50
Benzo(ghi)perylene	50
Chrysene	100
Naphtalene	50
Σ PAHs	1,000
Chlorinated hydrocarbons	
PCB Σ 7 (congeners)[1)]	20
PCDDs/Fs[2)]	1
HCB	20
DDT	30
DDE	25
DDD	20
HCH (Σ α+β+γ)	10
Non polar hydrocarbons	
hydrocarbons C_{10}-C_{40} (mg/kg)	100

Table 2. Proposed preventive limit values of persistent organic pollutants in agricultural soils of Czech Republic [1)] 28, 52, 101, 118, 138, 153, 180 [2)] value of I-TEQ PCDD/F (ng/kg)

District Teplice is situated in south-western direction from the Usti nad Labem district and its area is 469.27km^2. In the district there live 128,464 inhabitants in 34 municipalities,10 of which are classified as the towns. The urbanization rate reaches 84.08% of inhabitants. The capital of district is the town Teplice. The density of population has value of 274 inhabitants/km^2 in the district. The agricultural land covers 34.25% of district area and the ratio of arable land is 52.01%, it means 17.81% of district area. The other land covers 65.75% of district area and the ratio of forest is 55.91%, this is 36.76% of district area. There are two geomorphologic formations in district area, Central Bohemian Uplands (south-western and western part) and the Ore Mountains (northern part along the border with Germany).

District Most is situated in South West direction from Teplice district and its area is 467.16km^2. In the district live 114 795 inhabitants in 26 municipalities 6 of which are classified as towns. 88.71% of inhabitants live in the towns. The capital of district is Most. The density of population has value of 246 inhabitants/km^2 in the district.

The agricultural land covers 29.27% of district area and the ratio of arable land is 69.98%, it means 20.48% of district area. The other land covers 70.73% of district area and the ratio of forest is 46.85%, this is 33.14% of district area. There are two geomorphologic formations in the district area, the Central Bohemian Uplands (south-western and western part) and the Ore Mountains (northern part on the border with Germany). The coal-field area is situated under the Ore Mountaims and the active open mines still exist in south-western part of the region. Large opencast mine closed in the 80s of 20th century is spread close to the Most town on the area of the former Most old town (destroyed before mining). This land is under reclamation in present time (artificial lake).

POPs	Indication value (mg/kg of dry matter)
Benzo(a) pyrene	2.0
sum PAHs [1)]	30.0
sum PCB [2)]	1.0
DDT and metabolites	4.0
HCH (α, β, γ)	0.1
HCB	0.1
PCDDs/Fs [3)]	20.0
Benzene	0.5
Ethylbenzene	5.0
Toluene	10.0
Xylene	10.0
hydrocarbons C_{10}-C_{40}	500

Table 3. Proposed indication limit values of persistent organic pollutants in Czech agricultural soils [1)] The sum of 16 individual PAHs (EPA) [2)] The sum of 7 PCB congeners (28+52+101+118 +138+153+180) [3)] ng/kg I-TEQ PCDDs/Fs

District Chomutov is situated in south-western direction from the Most district and its area is 935.3km^2. In the district live 125,758 inhabitants in 44 villages, 8 of which are classified as towns. 86.43% of inhabitants live in towns. The capital of district is Chomutov. The density of population has value of 134 inhabitants/km^2 in the district. The agricultural land covers 41.93% of district area and the ratio of arable land is 60.72%, it means 25.46% of district area.

The other land covers 58.07% of district area and the ratio of forest is 63.41%, this is 36.82% of district area. There are 2 geomorphologic formations in district area, Central Bohemian Uplands (south-western and western part) and the Ore Mountains (North part on the border with Germany).

2.1.1.2. North Bohemian coal field

Tectonic depression takes the area of the districts Teplice, Most and Chomutov (Figure 2). The coalfield is the relict of the Tertiary sedimentary basin filled in Miocene [13]. The layer of clay, sand and organic materials of 500 m thickness was lodged in the period of 17-22 millions years ago. The brown coal bed was developed from the peat layers laid in Tertiary marsh on the majority of basin area. The sedimentation of clay and sand prevailed in the estuary areas of rivers entering the marsh [14]. The bed is filled by river or delta sediments in these places completely. The brown coal bed was developed relatively evenly in the thickness of 25 – 45 m in the other part of the basin. The area of current rests of brown coal field has 870 km^2. The average altitude is 272 m.a.s. l.

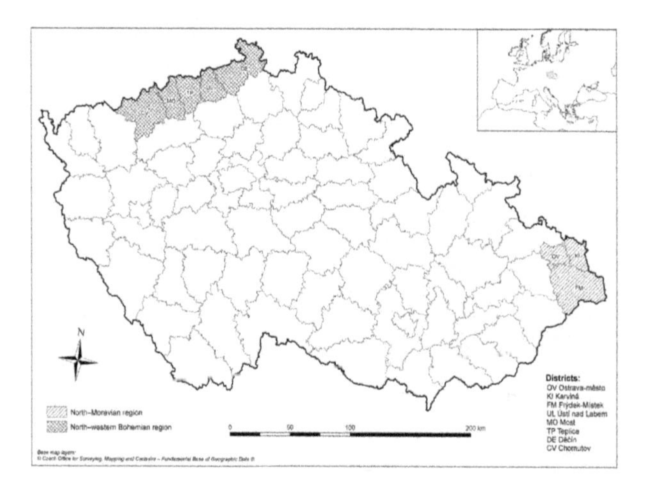

Figure 1. The area of North Bohemian Region and North Moravian Region in the Czech Republic

The mining activities have been running since 19th century and changed originally flat or downs surface relief. The process was accelerated after 1948 when the opencast mining technology on wide areas was elected. The mining is still continuing in opencast mines Bilina and Libus. The mine Bilina is the deepest mine of North Bohemian coal field with the depth of 200 m (the lowest point has the altitude of 35 m.a.s. l.). The total mining reserves were estimated at 165 millions tons in the beginning of the year 2012. The average content of ash is 26.9% and of total sulphur is 1.03% of dry matter of the coal. The open mine Libus is the largest mine of North Bohemian coal field and the total mining reserved were estimated at

240 millions tons in the beginning of the year 2012. The average content of ash is 36.8% and of total sulphur is 2.7% of dry matter of the coal. The coal is used for the production of energetic combustible mixtures. The environment is under influence of petrochemical industry in the Most district where the factory is located in Zaluzi u Mostu.

The North Bohemian coal field is impaired by strong anthropogenic activity when mining pits and mining depressions filled by water mining wastes in the form of table humps can be seen. The reclamations are done after landscape devastation.

2.1.1.3. The environmental data on the North Bohemian region

The area can be classified as the zone with high density of population and high concentration of the industry with increased level of imission pollutants. Following the information of Czech Hydrometeorological Institute [15] the concentration of dusty aerosol particles under 10μm (PM_{10}) were monitored on 27 localities in the region in 2009. The exceeding of 24 hours limit was observed on 7 localities. The maximum value was 63times higher than a daily limit value of 50μg/m^3. Nevertheless, there was oberverd no exceeding of year limit value in the region.

The values of dusty aerosol particles under 2,5μm ($PM_{2,5}$) were monitored on 6 localities in region. The highest value of annual concentration 19μg/m^3 is under limit value of European Direction 2008/50/EC [16].

The concentration of benzo(a)pyrene in the air was observed on 5 localities in the region and only 1 exceeding of target imission limit for annual concentration was detected in urban area of the city Usti nad Labem.

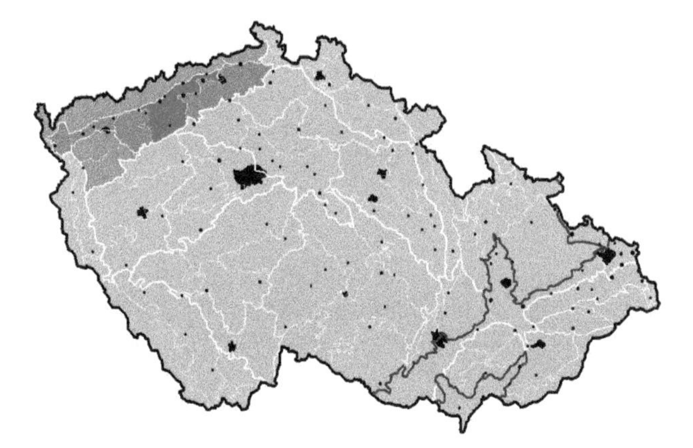

Figure 2. The North Bohemian coal field. The area in red. (the source: http://cs.wikipedia.org/wiki/Mosteck%C3%A1_p%C3%A1nev)

2.1.2. North Moravian region

2.1.2.1. Districts

The region has 3 districts: Ostrava, Karvina and Frydek-Mistek.

District Ostrava is situated in north-western part of the region and has the area of 331.53km^2. In the district 329,961 inhabitants live in 13 municipalities, 4 of which are classified as towns. The urbanization rate stands at 95%, indicating agglomeration character of the region. The capital of district is the Ostrava city. The density of population is 995 inhabitants/km^2 in the district.

The agricultural land covers 40.17% of district area and the ratio of arable land is 62.72%, it means 25.19% of district area. The other land covers 59.83% of district area and the ratio of forest is 18.18%, this is 10.88% of district area. The Ostrava-Karvina coal field is the part of geomorphologic formation shaped by the Ostrava basin a Karvina basin.

District Karvina is situated in north-eastern part of the region and its area is 356.24km^2. In the district 263,075 inhabitants live in 17 municipalities, 7 of which are classified as towns. 88.93% of inhabitants live in the towns. The capital of district is Karvina. The density of population has value 738 inhabitants/km^2 in the district. The agricultural land covers 50.77% of district area and the ratio of arable land is 68.82%, it means 34.94% of district area. The other land covers 49.23% of district area and the ratio of forest is 28.56%, this is 14.06% of district area. The district relief is shaped by geomorphologic formation of the Karvina basin as the part of the Ostrava-Karvina coal field.

District Frydek-Mistek is situated in south-eastern part of the region and its area is 1 208.49km^2. In the district 211,853 inhabitants live in 72 municipalities, 6 of which are classified as towns. The urbanization rate amount to 57.2% of inhabitants living in towns. The capital of district is Frydek-Mistek. The density of population has valuereaches 175 inhabitants/km^2 in the district. The agricultural land covers 39.38% of district area and the ratio of arable land is 48.95%, it means 19.28% of district area. The other land covers 60.62% of district area and the ratio of forest is 81.43%, this is 49.36% of district area. There are two important geomorphologic formations, the Moravskoslezske Beskydy Mountains and the Beskydy basin situated under mountainous area.

2.1.2.2. Ostrava-Karvina coal field

The Ostrava-Karvina coal field is Czech part of so called Upper Silesian coal pan spread on the area of Poland and partially of the Czech Republic that comprises coal fields in the Karvina, Ostrava and Beskydy basins. Ostrava-Karvina coal field is the largest coal field in the Czech Republic where black coal is extracted by deep mining technology. The Ostrava-Karvina coal field is separated in south part from the Beskydy basin by the Bludovicky break and has two parts (the Ostrava and Karvina basin) divided by the Orlová fault structure.

More than 300 km^2 is used for mining in Czech part of the Upper Silesian field and next 400 km^2 are considered to be perspective. The coal bearing layers in the rest of field area are not situated in accessible mining depths. Majority of mining activities runs in the Ostrava-Karvina coal field and only the Paskov mineone mine (Paskov) is open in the Beskydy coal field. There were udentified has two major coal layers in the Ostrava-Karvina coal field: Ostrava with general thickness of 2,880 m and Karvina with general thickness of 1,200 m. The average bed thickness of the Ostrava layer is 73 cm, whereas the most average thickness has the coal bed Prokop (2-4 m) with maximum more than 12 m. The average coal seam thickness of the Karvina layer is 180 cm and the most average thickness (504 cm) has also the coal bed Prokop with maximum up to 15m. The total amount of already extracted coal in the Ostrava-Karvina coal field is estimated at about 1,7 billions tons. In Czech part of the Upper Silesian field operate four deep mines and one mine stays in preserved regime at present time.

The mine CSM has the total area of 22.12km^2 and estimated activity will run to 2028.

The mine Karvina has the total area of 32.21km^2.

The mine Darkov has the total area of 25.9km^2.

The mine Paskov has the total area of 105.68km^2 including preserved mine Frenstat (63.17km^2).

The North Bohemian coal field is impaired by strong anthropogenic activity, especially the land subsidence is one of the most common and risky effects of the mining industry having impact on the regional environment. Other environmental hazards are connected to remaining existence of lagoon of by-products after black coal processing. The reclamations are running in the region at present time.

2.1.2.3. The environmental data on North Moravian region

The region can be characterised as the zone with high density of population (especially in the Ostrava and Karvina aglomeration) and with the presence of industrial activities mainly metallurgy. Following the information of Czech Hydrometeorological Institute [15] the concentration of dusty aerosol particles under 10μm (PM_{10}) in the region is the highest in the Czech Republic. The exceeding of limit values was detected on most of measured localities in the Ostrava and Karvina districts. It was observed that the concentrations of pollutants in the air rapidly increase during cold period of year influencing annual average value of pollution. The daily limit of PM_{10} concentration was exceeded more than 100 days a year on the most loaded localities in the Ostrava district. The target value for annual average concentration of dusty aerosol particles under 2.5μm ($PM_{2.5}$) given by European Directive 2008/50/ES (25μg/m^3) was exceeded on all observed localities in the North Bohemian region in 2009.

The limit concentration of benzo(a)pyrene (1ng/m^3) is permanently exceeded on most area of the North Bohemian region and multiple exceeding were observed on most localities. The maximum was detected in the Ostrava region (nine multiple of the limit).

2.2. Terrain works methodology

The plan of soil sampling was done first using map sources. The systematic soil sampling scheme based on the equidistance net of sampling points was prepared to maximally maintain the regular character of sampling. However the study targeted only agricultural soils and thus forest soils, urban soils and mining areas were excluded from sampling plan. The numbers of samples in individual districts of the North Bohemian and North Moravian imission regions are presented in table 4. The samples were taken out in the period of 2000 – 2005. The intensity of sampling was, on average, 1 sample/km^2 depending on the area of the district, number of samples and presence of forest and urban soils and mining areas. The potential sources of contamination were determined (industrial zones, mining zones, power stations) as well as the base meteorological data.

The soil samples were taken out from humic horizons of agricultural soils. The depth of the soil layer for sampling was between 5 and 15 cm. Each sample was collected from 10 partial samples on the locality. The sampling was done in minimal distance of 50 m from road. The samples were stored and transported in jars and frozen by the temperature –18 ^{0}C after the transport. Every locality was described and geographic coordinates were assessed using GPS. The determined soil characteristics, including soil type, soil subtype and soil sort (soil texture), pH value [17] and the content of C_{org} [1518], were compared with the contents of POPs. The POPs analysis was realised in accredited commercial laboratories.

North Bohemian imission region					North Moravian imission region		
179					106		
Decin	Usti n/L	Teplice	Most	Chomutov	Ostrava	Karvina	Frydek-Mistek
27	33	47	39	33	40	33	33

Table 4. The numbers of soil samples taken out in the North Bohemian and North Moravian regions and in individual districts.

2.3. Persistent organic pollutants analysis

The methodology of POPs analysis is following for individual groups.

BTEX (benzene, e-benzene, toluene and xylene)

Method used: EPA Method 8260 B [19]

Equipment: Gas chromatograph with mass spectrometer (GC/MS)

Principle

The soil samples are extracted by methanol and defined volume of extract is dosed into redistilled water after 24 hours. The final solution is analyzed by the system GC/MS using Headspace dozer. The individual substances are defined on the base of comparison of retention time and mass spectrum of analyzed substance considering mass spectrum of standard.

PAHs – polycyclic aromatic hydrocarbons

Method used: methodology TNV 75 8055 [20]

Equipment: High-performance liquid chromatograph with fluorescence detector (HPLC)

Principle

The solid sample analyse comprises the exsiccation using waterless sulphate and the extraction procedure in the acetone solution. The raw extract is analysed without purifying. PAHs are determined by the high-performance liquid chromatography with fluorescence detection (mobile phase – acetonitrile/water). One instrument measured some PAHs portions during isocratic elution under invariable wavelength and the second plant detected the other PAHs portion on equal terms. Two detectors in series assemble the instrument configuration thus two different wavelengths are involved in the detection. Such procedure minimises the difficulties with the gradient elution and with the alteration of the wavelength setting during analyse by the division of unpurified samples.

The concentration levels of individual compounds, the sum values of the compounds (the PAHs sum), the sum value of 2-3 nuclei PAHs and of 4-6 nuclei PAHs were used for the assessment of the load of soils and plants. The sum of toxic equivalency factors for PAHs (the TEF PAHs sum) was involved as well to take into account various toxicological characteristics of individual PAHs compounds. The TEF PAHs sum is defined as the sum of the products of the concentration of each compound multiplied by the toxic equivalent value for carcinogenic compounds. There were used following compounds:

Benzo(a)pyrene and Dibenzo(a,h)anthracene - toxic equivalent value = 1

Benzo(a)anthracene, Benzo(b)fluoranthene and Indeno(1,2,3-cd)pyrene - toxic equivalent value = 0,1

Benzo(k)fluoranthene - toxic equivalent value = 0,01

PCB7 – polychlorinated biphenyls, seven indicator congeners (28, 52, 101, 118, 138, 153, 180)

Method used: EPA Method 8082 [21]

Equipment: Gas chromatography with ECD detector (GC/ECD)

Principle

Dry soil sample is extracted by n-hexan. Extract is dozed after re-cleaning into gas chromatograph where the separation on capillary column and the detection on ECD detector are processed. The software identifies individual congeners on the base of comparison of retention times in calibrated solutions and samples.

DDT sum – sum of DDT, DDE and DDD

Method used: EPA Method 8082

Equipment: Gas chromatography with ECD detector (GC/ECD)

Principle

The dry sample is extracted by dichlormethane using intensive shaking and after volatilization is transferred into n-hexan. 1μl of the extract is dozed into gas chromatograph where the separation on capillary column and the detection on ECD detector are processed. The software identifies individual substances on the base of comparison of retention times in calibrated solutions and samples.

2.4. Results evaluation

The data were processed by the use of elementary statistical methods (Microsoft Excel) and geographic information systems (ESRI ArcGIS 9.2) was used for visualisation of soil load by sum of PAHs and benzo(a)pyrene in agricultural soils. The map outputs are based on the existence of contour lines connecting the points with identical concentration of observed substances. The Inverse Distance Weighting function was used for spatial interpolation. The areas of graduated concentrations connected with the other map layers (base geographical map etc.) create the map output.

3. Results and discussion

3.1. Soil load by polycyclic aromatic hydrocarbons (PAHs)

The comparison of regional load by PAHs shows very clear conclusion: The load of the North Moravian region by PAHs is demonstrably higher. The fundamental statistical data of soil load by sum of PAHs show table 5 (North Bohemian region) and table 6 (North Moravian region). The most loaded district is the Ostrava district with longterm metallurgical tradition where 85% of soil samples overcome preventive limit (1mg/kg) for sum of PAHs in Czech agricultural soils (reflecting background value). The observed maximal values differ ten times between the regions. The differences could be also found from the viewpoint of structural characteristics of contamination. While the participation of toxic beno(a)pyrene in total load by PAHs reaches about 7% in the North Bohemian region, its participation in the load reaches almost 17% in the North Moravian region (table 7). The load of individual district by PAHs can be documented by the comparison of number of exceeding of proposed indication limit for benzo(a)pyrene and sum of PAHs [12]. There was observed only one limit overrun among the North Bohemian samples, the exceeding was distinctive for benzo(a)pyrene and sum of PAHs (table 8). In the North Moravian region (table 9), there were documented seven localities exceeding indication limit for benzo(a)pyrene or sum of PAHs. Only two localities show exceeding for both indicators and maximum observed value is generally ten times higher than in the North Bohemian region. On the other hand must be accepted that soil load by PAHs was caused by floods on three localities (Stara Karvina) in the North Moravian region. Nevertheless, the problems with increased load by carcinogenic benzo(a)pyrene in the North Moravian region also follow from this comparison.

District	Samples number	Samples number increased	Samples increased %	Maximum value (mg/kg)
Decin	27	2	7.4	5.71
Usti/Labem	33	4	12.1	5.68
Teplice	47	11	23.4	7.67
Most	39	7	18	46.52
Chomutov	33	3	9.1	3.88
Together	179	27	15.1	46.52

Table 5. The load of agricultural soils by sum of PAHs in North Bohemian Region

District	Samples number	Samples number increased	Samples increased %	Maximum value (mg/kg)
Ostrava	40	34	85	28.1
Karvina	33	15	45,5	37.81
Frydek-Mistek	33	14	42,4	336.2
Together	106	63	59,4	336.2

Table 6. The load of agricultural soils by sum of PAHs in North Moravian Region

	Benzo(a)pyrene			Fluoranthene		
	Samples number increased	Samples increased %	Maximum value (mg/kg)	Samples number increased	Samples increased %	Maximum value (mg/kg)
North Bohemia	13	7.3	8.39	30	16.8	5.43
North Moravia	53	50	32.5	52	49.1	68.3

Table 7. The load of soil by benzo(a)pyrene and fluoranthene in North Bohemian and North Moravian Regions

The stronger impact of load by PAHs on health problems of inhabitants in the Nord Moravian region was presented by some authors [4,5]. Spatial distribution of POPs in both region is presented by contour maps where the load by benzo(a)pyrene and fluoranthene (the most widespread PAH substance) is visualised by GIS tools. The load by benzo(a)pyrene in the North Bohemian and North Moravian region shows Figure 3. The load in the North Bohemian region is given mainly by point sources of contamination and just two points with higher contents of sum of PAHs were found, the first near the Decin town and the second one in north-western part of the region. The load in the North Moravian region by carcinogenic benzo(a)pyrene shows surface contamination mainly in the field area around the Ostrava city, increased load is obvious in the vicinity of the Trinec town (ironworks) and the load of region spreads to the Moravskoslezske Beskydy Mountains in the east of region. The similar trend can be seen in the case of fluoranthene in the North Moravian region, but more visible surface load can bee seen in the North Bohemian region in comparison with benzo(a)pyrene (Figure 4). Very probably, more massive load by fluoranthene, that is typical for burning processes, is given by fossil combustibles in towns more than by industrial activities in the North Bohemian region. Moreover, the comparison of layout of coal-fired power stations in the region of North Bohemia (black squares indicate the areas of coal-fired power stations) and the presence of areas with increased load by fluoranthene indicates important role of their activity. Only one coal-fired power station in the North Moravian region (Detmarovice) situated close to border with Poland does not influence the load of region by PAHs as shows the map.

The quality of the load is evaluated on the base of ratio of PAHs substances with 2 nuclei, 3-4 nuclei and 5-6 nuclei. Table 10 presents the values of the sum of medians of individual substances grouped on the base of nuclei number. This value indicates not only the quality of the load but also the quantity (cumulative effect of observed concentrations). Only the quality of the load is defined by the values of medians of the groups PAHs differentiated by nuclei numbers. The both characteristics demonstrate significantly increased load by more nuclei substances in the North Moravian region in comparison with the North Bohemian region and increased toxicity of the load in the North Moravian region by PAHs consecutively.

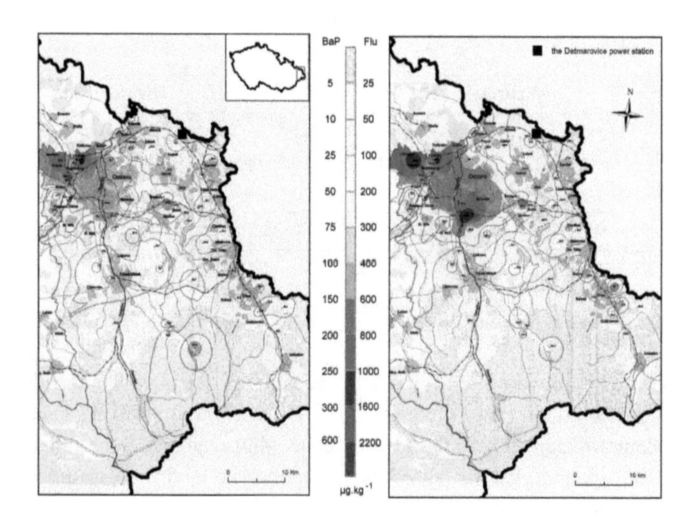

Figure 3. The load of the North Moravian Region by benzo(a)pyrene and fluoranthene (µg/kg)

The recent development in the North Moravian region proved no decreasing temporal tendency of air pollution by dust particles and also PAHs, because there spatially coincide various pollution sources in the region and the situation needs more complicated approach in comparison with the North Bohemian Region where technology improvement in coal-fired power stations influenced positively air quality.

Locality	District	Benzo(a)pyrene (mg/kg)	Sum of PAHs (mg/kg)
Horni Jiretin	Most	**8.39**	**46.52**

Table 8. The localities of the North Bohemian region with the load by PAHs exceeding proposed indication limit for benzo(a)pyrene and sum of PAHs in agricultural soil

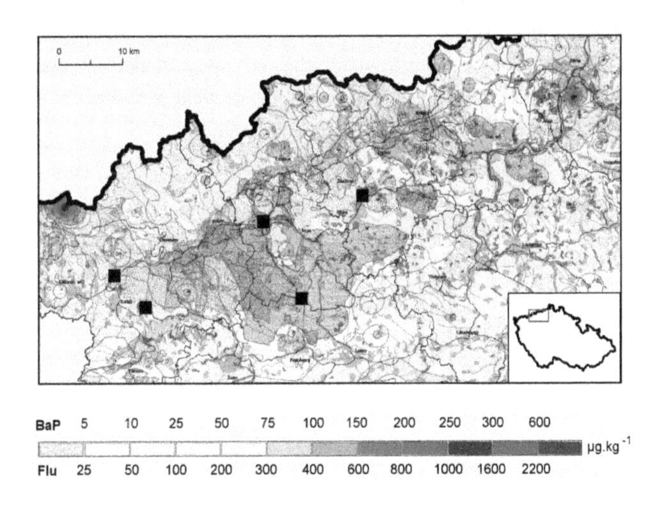

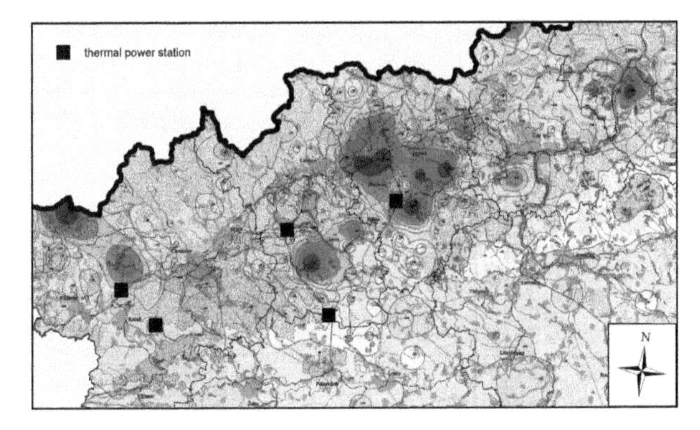

Figure 4. The load of the North Bohemian Region by benzo(a)pyrene and fluoranthene (μg/kg)

3.2. Soil load by monocyclic aromatic hydrocarbons (MAHs)

The soil load by individual substances (benzene, toluene, xylene, e-benzene) from the group of monocyclic aromatic hydrocarbons (MAHs) present tables 11 (North Bohemian Region) and 12 (North Moravian Region). The comparison shows that more loaded by MAHs is the North Bohemian region. The most loaded district is Most where petrochemical factory (Zaluzi u Mostu) is situated. The region Most shows not only the highest number of limit exceeding but also the highest maximal values. E-benzene is the substance with maximal detected value (0,9 mg/kg) at all. Opposite, no exceeding of e-benzene limit in the soil was observed in the North Moravian region where only very slight exceeding of preventive value of benzene, toluene and xylene was observed. Only exceeding of preventive limit was observed and no exceeding of indication limit was detected in the soils of both regions. It is evident that the load by monocyclic aromatic hydrocarbons immediately relates to chemical industry and there is no correlation with the load by polycyclic aromatic hydrocarbons connected with burn processes.

Locality	District	Benzo(a)pyrene (mg/kg)	Sum of PAHs (mg/kg)
Przno	Frydek-Mistek	**2.31**	23.12
Paskov I	Frydek-Mistek	**32.5**	**336.17**
Paskov II	Frydek-Mistek	0.19	**95.68**
Ostrava-Arnostovice	Ostrava	**2.98**	28.06
*Stare Mesto Karvina I	Karvina	**2.72**	25.01
*Stare Mesto Karvina II	Karvina	**3.18**	27.89
*Stare Mesto Karvina III	Karvina	**3.56**	**37.81**

Table 9. The localities of the North Moravian region with the load by PAHs exceeding proposed indication limit for benzo(a)pyrene and sum of PAHs in agricultural soil * The load by PAHs after floods – fluvial load

	North Bohemian Region			North Moravian Region		
	2n	3-4n	5-6n	2n	3-4n	5-6n
Sum of substance. medians (mg/kg)	0.002	0.192	0.018	0.0006	1.337	0.173
Ratio	1	92.6	9	1	2227.8	288.2
Medians of n-groups (mg/kg)	0.002	0.01	0.008	0.0006	0.153	0.0437
Ratio	1	5	4	1	255.1	72.8

Table 10. The sum of medians and medians of PAHs substances in soil differentiated by nuclei number

District	Samples number	Samples number increased	Samples increased %	Maximum value (mg/kg)
Decin	27	0	0	0.02 e-benzene
Usti/Labem	33	0	0	0.04 e-benzene
Teplice	47	1 benzene 1 toluene 4 xylene 6 e-benzene	2.13 benzene 2.13 toluene 8.51 xylene 12.77 e-benzene	0.08 benzene 0.08 toluene 0.12 xylene 0.1 e-benzene
Most	39	3 benzene 3 toluene 6 xylene 6 e-benzene	7.7 benzene 7.7 toluene 15.4 xylene 15.4 e-benzene	0.11 benzene 0.15 toluene 0.48 xylene 0.90 e-benzene
Chomutov	33	0	0	0.04 e-benzene
Together	179	4 benzene 4 toluene 10 xylene 12 e-benzene	2.23 benzene 2.23 toluene 5.59 xylene 6.70 e-benzene	0.11 benzene 0.15 toluene 0.48 xylene 0.90 e-benzene

Table 11. The load of soils by individual monocyclic aromatic hydrocarbons in the North Bohemian region

3.3. The load of soils by chlorinated hydrocarbons (PCBs, DDTs)

The soil load by sum of seven PCB congeners (28, 52, 101, 118, 138, 153, 180) in the North Bohemian and North Moravian region are presented in tables 13 and 14. There is distinctive difference between the loads of the regions when the North Moravian region is loaded much more. More than 25% samples exceeded proposed preventive limit for PCB7 in agricultural soils in the North Moravian region. Only 3.9% samples exceeded preventive limit in the North Bohemian region. The most loaded district from both regions is the district Ostrava where 42.5% samples exceeded preventive limit. The maximal value was detected in the sample from the district Decin in the North Bohemian region. The value 0.52 mg/kg is 26 multiple of proposed preventive limit (0.02 mg/kg) and about one half of proposed indication limit (1 mg/kg) for PCB7 in the soil. The sources of PCB in the environment are concentrated in industrial areas and usually relate to the presence of dumps and waste incinerators [22]. PCB load may relate to historical environmental burdens when PCB revolatilization from soil depends on weather conditions. These processes very probably reflect the soil load of the Ostrava district with high concentration of human activities. The increased soil load was for example confirmed in the urban area of the Prague city by our investigation and general load of Czech agricultural soils moves in the interval of 1.19 – 20.11 μg/kg (geometric means calculated for individual districts of the Czech Republic). The average PCB7 contents in humic horizons of Czech agricultural soils ranged between 5.5 μg/kg (2000 – 2003) and 8.43 μg/kg (2004) following the data of Czech National Stocktaking [23].

District	Samples number	Samples number increased	Samples increased %	Maximum value (mg/kg)
Ostrava	40	0 benzene	0	0.017 benzene
		1 toluene	2.5 toluene	0.031 toluene
		1 xylene	2.5 xylene	0.039 xylene
		0 e-benzene	0 e-benzene	0.009 e-benzene
Karvina	33	1 benzene	3.03 benzene	0.032 benzene
		0 toluene	0 toluene	0.010 toluene
		0 xylene	0 xylene	0.009 xylene
		0 e-benzene	0 e-benzene	0.004 e-benzene
Frydek-Mistek	33	0 benzene	0 benzene	0.020 benzene
		0 toluene	0 toluene	0.003 toluene
		0 xylene	0 xylene	0.010 xylene
		0 e-benzene	0 e-benzene	0.002 e-benzene
Together	106	1 benzene	0.94 benzene	0.032 benzene
		1 toluene	0.94 toluene	0.031 toluene
		1 xylene	0.94 xylene	0.039 xylene
		0 e-benzene	0 e-benzene	0.009 e-benzene

Table 12. The load of soils by individual monocyclic aromatic hydrocarbons in the North Moravian region

District	Samples number	Samples number increased	Samples increased %	Maximum value (mg/kg)
Decin	27	4	14.82	0.52
Usti/Labem	33	2	6.06	0.08
Teplice	47	0	0	0.01
Most	39	1	2.56	0.05
Chomutov	33	0	0	0.01
Together	179	7	3.91	0.52

Table 13. The load of soils by sum of PCB_7 in the North Bohemian region

District	Samples number	Samples number increased	Samples increased %	Maximum value (mg/kg)
Ostrava	40	17	42.5	0.43
Karvina	33	5	15.15	0.13
Frydek-Mistek	33	5	15.15	0.26
Together	106	27	25.47	0.43

Table 14. The load of soils by sum of PCB_7 in the North Moravian region

District	Samples number	Samples number increased DDT/DDE/DDD	Samples increased % DDT/DDE/DDD	Maximum value (mg/kg) DDT/DDE/DDD
Decin	27	0/0/0	0/0/0	0.001/0.001/0.001
Usti/Labem	33	1/0/0	3.03/0/0	1.13/0.001/0.001
Teplice	47	4/4/3	8.5/8.5/6.4	1.21/0.15/0.26
Most	39	6/1/0	15.4/2.6/0	0.06/0.03/0.01
Chomutov	33	1/0/0	3.03/0/0	0.04/0.001/0.001
Together	179	12/5/3	6.7/2.8/1.7	1.21/0.15/0.26

Table 15. The load of soils by DDT and its metabolites in the North Bohemian region

DDT and its congeners have been still registered in agricultural soil of the Czech Republic in spite of DDT use finished in 1974 [22]. DDT decomposition is generally reflected by slowly increased contents of DDE in soils. This trend is more visible in the North Moravian region where the ratio of DDT/DDE/DDD has value 3.5/4/1 while in the North Bohemian region has value 4/1.7/1 (indicating relatively more recent load). We have the hypothesis about non legal use of high doses of DDT in the North Bohemian forests (the Ore Mountains) during the bark beetle calamity in the eighties of 20th century (unsworn information). The level of soil load by DDT and its metabolites is comparable in both regions. Only the Decin district in the North Bohemian region, where no exceeding was observed, differs from the other districts. This fact shows no gap with our hypothesis because the Ore Mountains range does not reach the Decin district area. The general trends of DDTs concentration are complicated in the soils of the Czech Republic because a strong oscillation of DDTs values was observed [23]. Nevertheless the data of Czech National Stocktaking show the decreasing number of DDT exceeding of preventive limit during 2000 – 2002 years (from 60 to 18) and increasing number of DDE exceeding of preventive limit (from 14 to 24).

Only the maximal values of DDTs in the soils of both regions exceed preventive limits and the maximum 1.21 mg/kg reaches 30.25% of proposed indication limit value.

District	Samples number	Samples number increased DDT/DDE/DDD	Samples increased % DDT/DDE/DDD	Maximum value (mg/kg) DDT/DDE/DDD
Ostrava	40	3/3/0	7.5/7.5/0	0.35/0.18/0.005
Karvina	33	2/4/1	6.01/12.1/3.03	0.069/0.032/0.049
Frydek-Mistek	33	2/1/1	6.01/3.03/3.03	0.057/0.033/0.022
Together	106	7/8/2	6.6/7.55/1.89	0.35/0.18/0.049

Table 16. The load of soils by DDT and its metabolites in the North Moravian region

4. Conclusion

The data of soil load by observed POPs groups in two environmentally affected areas indicate generally higher load of the North Moravian region in comparison with the North Bohemian region. This result is especially supported by the comparison of soil load by polycyclic aromatic hydrocarbons. The load relate to spatial coincidence of various pollution sources connected with high concentration of metallurgy and with high urbanization rate in the Ostrava agglomeration. The load of soil by polyaromatic hydrocarbons has diffuse character and exceeding of proposed preventive limit value (based on PAHs background values in Czech soils) was detected on most observed localities. The exceeding of proposed indication limit value for PAHs in soil (derived from human health risks) was detected in the North Moravian region as well. Markedly increased soil load was also monitored in the case of PCBs in the North Moravian region where the effect of increased urban environment plays important role. Nevertheless, the intensity of the load by PCBs is lower and only preventive limit values were exceeded. The different trends were proved for two POPs groups – monoaromatic hydrocarbons and DDTs and its metabolite. While soil load by DDTs is comparable in both regions (with some qualitative differences) the load of agricultural soils by MAHs is markedly higher in the North Bohemian region and especially in the Most district. This is the consequence of petrochemical industry situated close to the Most town. The recent development in both regions may be evaluated as better in the North Bohemian region where the investment into technologies in coal-fired power stations decreased the load of environment by contaminants. The adverse emission situation in the North Moravian region with increased concentration of metallurgy is still remaining.

5. Lists of abbreviations

C_{org}-Content of organic carbon

DDT-Dichlorodiphenyltrichloroethane

MAHs-Monoaromatic hydrocarbons

PAHs-Polyaromatic hydrocarbons

PCB7-Sum of 7 polychlotinated biphenyls congeners

PCDDs/Fs-Polychlorinated dibenzodioxins/furans

PM_{10}-Particulate matter in air of size <10 μm

$PM_{2.5}$-Particulate matter in air of size < 2.5 μm

POPs-Persistent organic pollutants

REs-Risky elements

Acknowledgements

The chapter was prepared by the support of the Project of Ministry of Agriculture MZE0002704902.

Author details

Radim Vácha*, Jan Skála, Jarmila Čechmánková and Viera Horváthová

*Address all correspondence to: vacha@vumop.cz

Research Institute for Soil and Water Conservation, Prague, Czech Republic

References

[1] Podlešáková, E. (1992). The pollution of atmosphere. *The influence of Agriculture on Environment, Čihalík J. ed. Ministry of Agriculture of Czech Republic, Brazda press*, 32-37.

[2] Bouska, V., & Pesek, J. (1999). Quality parameters of lignite of the North Bohemian Basin in the Czech Republic in comparison with the world average lignite. *International Journal of Coal Geology*, 40(2), 211-235.

[3] CEZ Group. (2012). *Coal-fired Power Plants in CR*, http://www.cez.cz/en/power-plants-and-environment/coal-fired-power-plants/cr/chvaletice.html, accessed 17 August 2012.

[4] Rossnerova, A., Spatova, M., Rossner, P., Novakova, Z., Solansky, I., & Sram, R. J. (2011). Factors affecting the frequency of micronuclei in asthmatic and healthy children from Ostrava. *Mutation Research-Fundamental and Molecular Mechanisms of Mutagenesis*, 708(1-2), 44-49.

[5] Sram, R. J., Mitcova, A., Rossner, P., Rossnerova, A., Schmuczerova, J., Solansky, I., Spatova, M., Svecova, V., & Topinka, J. (2010). European Hot Spot of Air Pollution by PM2.5 and B[a]P, Ostrava, Czech Republic. *Environmental And Molecular Mutagenesis*, 51(7), 737.

[6] Podlešáková, E., Němeček, J., & Hálová, G. (1996). The proposal of soil contamination limits of potentially risky elements for CR. *Rostlinná výroba*, 42(3), 119-125.

[7] Němeček, J., Podlešáková, E., & Pastuszková, M. (1996). Proposal of soil contamination limits for persistent organic xenobiotic substances in the Czech Republic. *Rostlinna Vyroba*, 42(2), 49-53.

[8] EPA, United States Environmental Protection Agency. (2012). http://www.epa.gov/superfund/students/wastsite/soilspil.htm, accessed 17 August.

[9] European Parliament. Thematic Strategy for Soil Protection. *European Parliament resolution on the Commission communication'Towards a Thematic Strategy for Soil Protection' (COM 179- C5-0328/2002- 2002/2172(COS)).*

[10] Ministry of Environment of the Czech Republic. (1994). The notice of the Ministry of Environment for the management of the soil protection. *Coll* [13].

[11] Sáňka, M., Němeček, J., Podlešáková, E., Vácha, R., & Beneš, S. (2002). The elaboration of limit values of concentrations of risky elements and organic persistent compounds in the soil and their uptake by plants from the viewpoint of the protection of plant production quantity and quality. *The report of the Ministry of Environment of the Czech Republic*, 60.

[12] Sáňka, M., & Vácha, R. (2006). The evaluation of the limits for soils in normative regulations of CR and selected European countries. *The report of the Ministry of Environment of the Czech Republic*, 38.

[13] Kvaček, Z., Böhme, M., Dvořák, Z., Konzalová, M., Mach, K., Prokop, J., & Rajchl, M. (2001). Early Miocene freshwater and swamp ecosystems of the Most Basin (north Bohemia) with particular reference to the Bílina Mine section. *Journal of the Czech Geological Society*, 49-1.

[14] Dvořák, Z., & Mach, K. (1999). Deltaic deposits in the North-Bohemian brown coal basin and their documentation in the Bílina opencast mine. *Acta Universitatis Carolinae, Geologica*, 43(4), 633-641.

[15] CHMI, Czech Hydrometeorological Institute. (2012). http://portal.chmi.cz/files/portal/docs/uoco/isko/grafroc/groc/gr09cz/kap.241.html, accessed 18 August.

[16] Directive 2008/50/EC of The European Parliament and of The Council of May of ambient air quality and cleaner air for Europe. http://eurlex.europa.eu/LexUriServ/LexUriServ.do?uri=OJ:L:2008:152:0001:0044:EN:PDF, accessed 20 August 2012.

[17] ČSN ISO 10390. (1996). Soil Quality- Determination of pH. *Czech Normalisation Institute Prague*, 12.

[18] Zbíral, J., Honsa, I., Malý, S., & Čižmár, D. (2004). Soil Analysis III. *Central Institute for Supervising ans Testing in Agriculture Brno*, 199.

[19] EPA Method 8260B. (2012). Volatile organic compounds by gas chromatography/Mass Spectroscopy (GS/MS). *US Environmental Agency (EPA), USA86*, http://www.epa.gov/osw/hazard/testmethods/sw846/pdfs/8260b.pdf, accessed 20 August.

[20] TNV 75 8055. (2012). Sludge Characterisation- The Analysis of Selected Polycyclic Aromatic Hydrocarbons (PAHs) by HPLC with fluorescence detection method. http://shop.normy.biz/d.php?k=165109, accessed 20 August.

[21] EPA Method 8082. (1996). Polychlorinated Biphenyls (PCBs) by Gas Chromatography. *US Environmental Agency (EPA), USA.*, http://www.caslab.com/EPA-Methods/PDF/EPA-Method-8082.pdf, accessed 20 August 2012.

[22] Asher, B. J., Ross, MS, & Wong, C. S. (2012). Tracking chiral polychlorinated biphenyl sources near a hazardous waste incinerator : Fresh emmissions or weathered revolatilization? *Environmnetal Toxicology and Chemistry*, 31(7), 1453-1460.

[23] Holoubek, I., Adamec, V., Bartoš, M., Černá, M., Čupr, P., Bláha, K., Bláha, L., Demnerová, K., Drápal, J., Hajšlová, J., Holoubková, I., Jech, L., Klánová, J., Kohoutek, J., Kužílek, V., Machálek, P., Matějů, V., Matoušek, J., Matoušek, M., Mejstřík, V., Novák, J., Ocelka, T., Pekárek, V., Petira, O., Punčochář, M., Rieder, M., Ruprich, J., Sáňka, M., Vácha, R., & Zbíral, J. (2003). National stocktaking of Persistent organic pollutants in the Czech Republic. *Project GF/CEH/01/003 Enabling activities to facilitate early action on the implementation of the Stockholm Convention on Persistent organic pollutants (POPs) in the Czech Republic. TOCOEN REPORT* [249], http://www.genasis.cz/stockholm-stockholmska_umluva-inventura_pops_2007/, accessed 22 August 2012.

Section 2

Risk

Chapter 3

What Do We Know About the Chronic and Mixture Toxicity of the Residues of Sulfonamides in the Environment?

Anna Białk-Bielińska, Jolanta Kumirska and Piotr Stepnowski

Additional information is available at the end of the chapter

1. Introduction

Thanks to their low cost and their broad spectrum of activity in preventing or treating bacterial infections, sulfonamides (SAs) are one of the oldest groups of veterinary chemotherapeutics, having been used for more than fifty years. To a lesser extent they are also applied in human medicine. After tetracyclines, they are the most commonly consumed veterinary antibiotics in the European Union. As these compounds are not completely metabolized, a high proportion of them are excreted unchanged in feces and urine. Therefore, both the unmetabolized antibiotics as well as their metabolites are released either directly to the environment in aquacultures and by grazing animals or indirectly during the application of manure or slurry [1-3].

Physico-chemical properties and chemical structures of selected SAs are presented in Table 1. They are fairly water-soluble polar compounds, the ionization of which depends on the matrix pH. All the sulfonamides, apart from sulfaguanidine, are compounds with two basic and one acidic functional group. The basic functional groups are the amine group of aniline (all the SAs) and the respective heterocyclic base, specific to each SA. The acidic functional group in the SAs is the sulfonamide group. With such an SA structure, these compounds may be described by the pK_{a1}, pK_{a2} and pK_{a3} values corresponding to the double protonated, once protonated and neutral forms of SA (Table 1) [3-7].

Substance [CAS] Abbreviation	**Chemical structure**	**Selected physico-chemical properties**
Sulfaguanidine [57-67-0] SGD	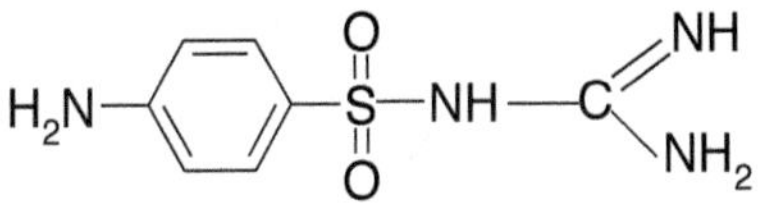 H_2N, O, S, O, NH, C, NH, NH_2	M = 214.2 g mol^{-1} pK_{a2} = 2.8 pK_{a3} = 12.0 logP = -1.22
Sulfapyridine [144-83-2] SPY	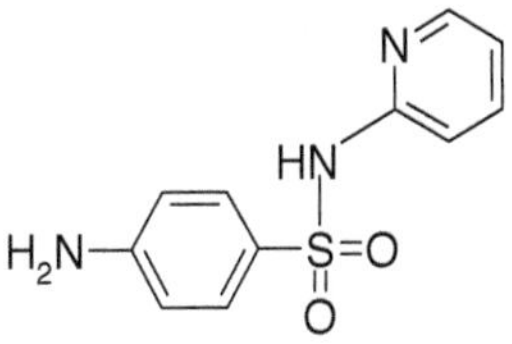 N, HN, H_2N, S=O, O	M = 249.2 g mol^{-1} pK_{a2} = 2.37 pK_{a3} = 7.48 logP = 0.03
Sulfadiazine [68-35-9] SDZ	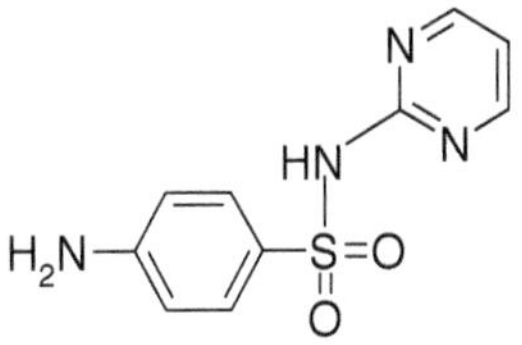 N, N, HN, H_2N, S=O, O	M = 250.3 g mol^{-1} pK_{a2} = 1.98 pK_{a3} = 6.01 logP = -0.09
Sulfamethoxazole [723-46-6] SMX	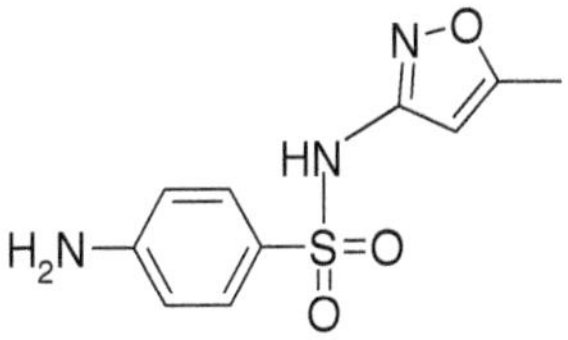 N-O, HN, H_2N, S=O, O	M = 253.3 g mol-1 pK_{a2} = 1.81 pK_{a3} = 5.46 logP = 0.89
Sulfathiazole [72-14-0] STZ	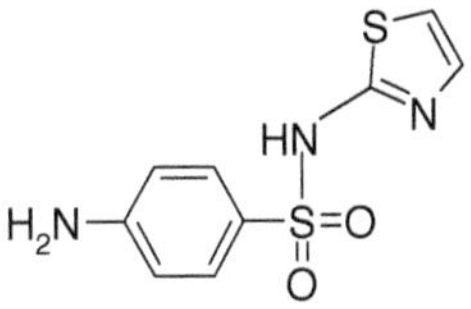 S, N, HN, H_2N, S=O, O	M = 255.3 g mol^{-1} pK_{a2} = 2.06 pK_{a3}= 7.07 logP = -0.04
Sulfamerazine [127-79-7] SMR	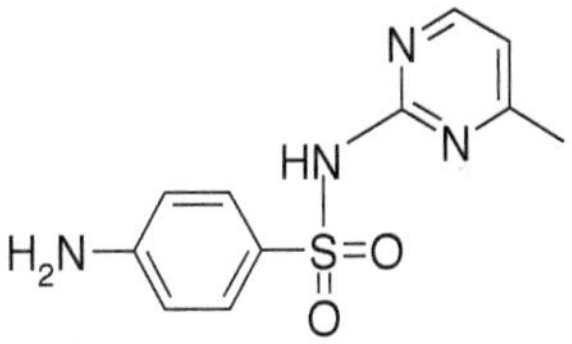 N, N, HN, H_2N, S=O, O	M = 264.3 g mol^{-1} pK_{a2} = 2.16 pK_{a3} = 6.80 logP = 0.11

Substance [CAS] Abbreviation	Chemical structure	Selected physico-chemical properties
Sulfisoxazole [127-69-5] SSX	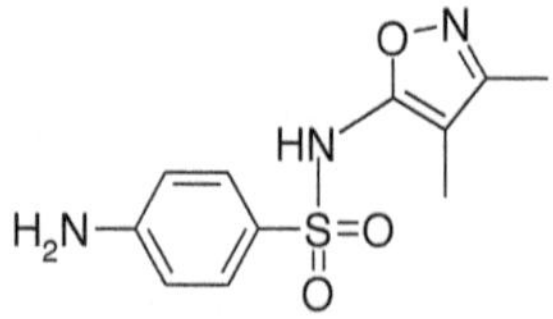	$M = 267.3\ g\ mol^{-1}$ $pK_{a2} = 2.15$ $pK_{a3} = 5.00$ $logP = 1.01$
Sulfamethiazole [144-82-1] SMTZ	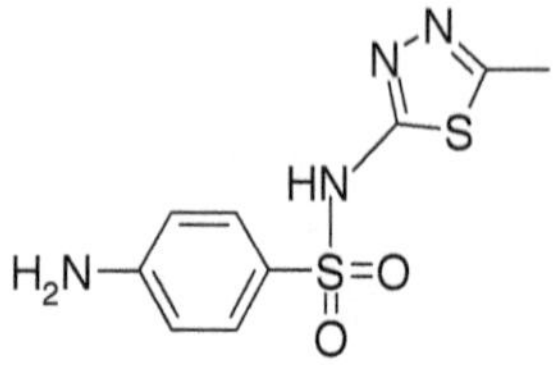	$M = 270.3\ g\ mol^{-1}$ $pK_{a2} = 2.24$ $pK_{a3} = 5.30$ $logP = 0.47$
Sulfadimidine (sulfamethazine) [57-68-1] SDMD (SMZ)	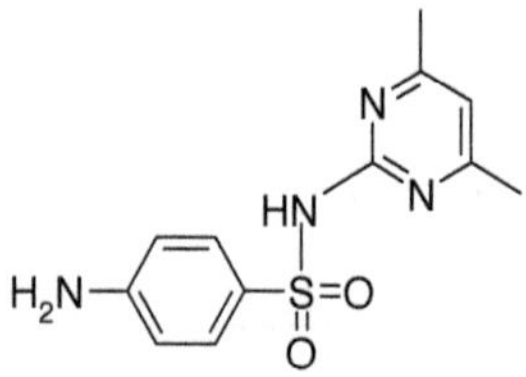	$M = 278.3\ g\ mol^{-1}$ $pK_{a2} = 2.46$ $pK_{a3} = 7.45$ $logP = 0.27$
Sulfamethoxypyridazine [80-35-3] SMP	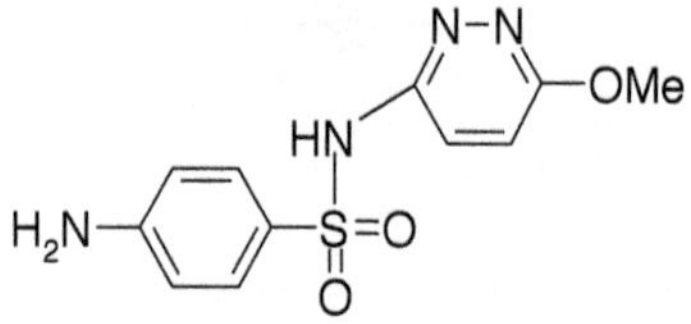	$M = 280.3\ g\ mol^{-1}$ $pK_{a2} = 2.20$ $pK_{a3} = 7.20$ $logP = 0.32$
Sulfachloropyridazine [80-32-0] SCP	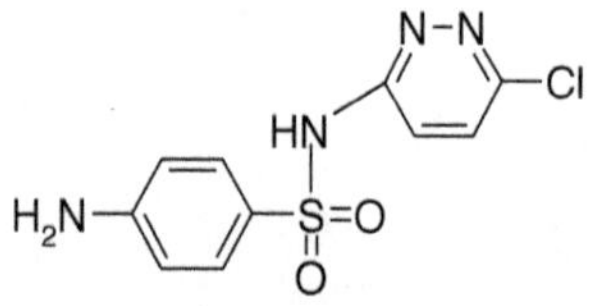	$M = 284.7\ g\ mol^{-1}$ $pK_{a2} = 1.72$ $pK_{a3} = 6.39$ $logP = 0.71$
Sulfadimethoxine [122-11-2] SDM	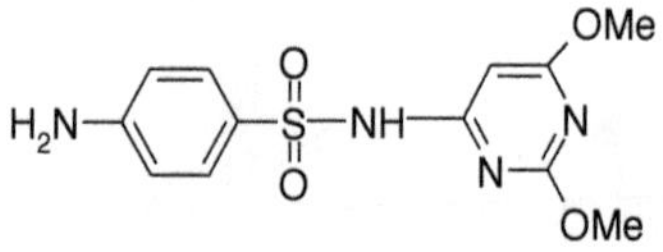	$M = 310.3\ g\ mol^{-1}$ $pK_{a2} = 2.5$ $pK_{a3} = 6.0$ $logP = 1.63$

Table 1. Structures and physico-chemical properties of selected sulfonamides (according to [1,6-13])

Due to their properties, after disposal in soils, these compounds may enter surface run-off or be leached into the groundwater. Moreover, they are also quite persistent, non-biodegradable and hydrolytically stable, which explains why in the last ten years they have been regularly detected not only in aquatic but also in terrestrial environments [1-3,7,14]. Although SAs concentrations in environmental samples are quite low (at the $\mu g\ L^{-1}$ or $ng\ L^{-1}$ level), they are continuously being released [3,15]. Therefore, the kind of exposure organisms may be subjected to will resemble that of traditional pollutants (e.g. pesticides, detergents), even those of limited persistence. Consequently, SAs as well as other pharmaceuticals may be considered pseudo-persistent.

SAs are designed to target specific metabolic pathways (they competitively inhibit the conversion of *p*-aminobenzoic acid, PABA) by inhibiting the biosynthetic pathway of folate (an essential molecule required by all living organisms), so they not only affect bacteria (target organisms) but can also have unknown effects on environmentally relevant non-target organisms, such as unicellular algae, invertebrates, fish and plants [16-18]. Belonging to different trophic levels, these taxonomic groups may be exposed to by SAs to various extents [15-16,19-20].

However, knowledge of the potential effects of SAs on the environment is very limited. Recently, a few review papers have been published that summarize the available ecotoxicity data of pharmaceuticals, including some sulfonamides [16-17,19-21]. Such data as are available on the potential effects of pharmaceuticals in the environment appear to indicate a possible negative impact on different ecosystems and imply a threat to public health. However, if we look just at the sulfonamides, most current studies have investigated acute effects mainly of single compounds and mostly with reference to sulfamethoxazole (SMX), one of the most common SAs, used in both veterinary and human medicine [16-17,20]. Available information on the ecotoxicity of selected sulfonamides has been review and is presented in Table 2.

Substance	Bacteria *Vibrio fischeri*	Green algae / Cyanobacteria/ Diatom*	Plants**	Invertebrates***		Vertebrates****
SGD	>50$_{\text{(30 min)}}$	43.56$_{\text{(96h, P. subcapitata)}}$	30.30$_{\text{(7d, L.gibba)}}$	0.87$_{\text{(48h, D. magna)}}$		
		3.40$_{\text{(96h, S. dimorphus}}$)	0.22$_{\text{(7d, L. minor)}}$			
		16.59$_{\text{(96h, S. leopoliensis)}}$				
		3.42$_{\text{(24h, S.vacuolatus)}}$				
SPY	>50$_{\text{(30 min)}}$	5.28$_{\text{(24h, S.vacuolatus)}}$	0.46$_{\text{(7d, L. minor)}}$			
SDZ	>25$_{\text{(30 min)}}$	7.80$_{\text{(72h, P. subcapitata)}}$	0.07$_{\text{(7d, L. minor)}}$	221$_{\text{(48h, D. magna)}}$		
		2.19$_{\text{(72h, P. subcapitata)}}$		13.7$_{\text{(21d, D. magna)}}$		
		0.135$_{\text{(72h, M. aeruginosa)}}$		212$_{\text{(48h, D. magna)}}$		
		2.22$_{\text{(24h, S.vacuolatus)}}$				
SMX	23.3$_{\text{(30 min)}}$	1.53$_{\text{(72h, P. subcapitata)}}$	0.081$_{\text{(7d, L.gibba)}}$	189.2$_{\text{(48h, D.magna)}}$	123.1$_{\text{(48h, D.magna)}}$	>750$_{\text{(48h, O. latipes)}}$ [a]
	>84$_{\text{(30 min)}}$	0.15$_{\text{(96h, P. subcapitata)}}$	0.132$_{\text{(7d, L.gibba)}}$	177.3$_{\text{(96h, D.magna)}}$	205.1$_{\text{(48h, D.magna)}}$	562.5$_{\text{(96h, O. latipes)}}$ [a]
	78.1$_{\text{(15 min)}}$	0.52$_{\text{(72h, P. subcapitata)}}$	0.0627$_{\text{(14d, D.carota)}}$	25.2$_{\text{(24h, D. magna)}}$	70.4$_{\text{(48h, M.macrocopa)}}$	27.36$_{\text{(24h, O. myskiss)}}$
	74.2$_{\text{(5 min)}}$	2.4$_{\text{(96h, C. meneghiniana)}}$	0.0612$_{\text{(21d, D.carota)}}$	15.51$_{\text{(48h, C. dubia)}}$	84.9$_{\text{(24h, M.macrocopa)}}$	
	>100$_{\text{(30 min)}}$	0.0268$_{\text{(96h, S. leopoliensis)}}$	0.0454$_{\text{(28d, D.carota)}}$	0.21$_{\text{(7d, C. dubia)}}$	9.63$_{\text{(48h, B.calyciflorus)}}$	

Substance	Bacteria *Vibrio fischeri*	Green algae / Cyanobacteria/ Diatom*	Plants**	Invertebrates***	Vertebrates****
		1.54$_{\text{(24h, S.vacuolatus)}}$	0.21$_{\text{(7d, L. minor)}}$	100$_{\text{(48h, C. dubia)}}$ 35.36$_{\text{(24h, T.platyurus)}}$^{a}	
STZ	>1000$_{\text{(15min)}}$ >50$_{\text{(30 min)}}$	13.10$_{\text{(24h, S.vacuolatus)}}$	3.552$_{\text{(7d, L.gibba)}}$ 4.89$_{\text{(7d, L. minor)}}$	149.3/85.4$_{\text{(48h/96h, D. magna)}}$ 616.7$_{\text{(24h, D. magna)}}$ 391/430$_{\text{(48h/24h, M. macrocopa)}}$ 135.7/78.9$_{\text{(48h/96h, D. magna)}}$	>500$_{\text{(48h, O. latipes)}}$^{a} >500$_{\text{(96h, O. latipes)}}$^{a} >100$_{\text{(48h, O. myskiss)}}$^{a}
SMR	>50$_{\text{(30 min)}}$	11.90$_{\text{(24h, S.vacuolatus)}}$	0.68$_{\text{(7d, L. minor)}}$		
SSX	>50$_{\text{(30 min)}}$	18.98$_{\text{(24h, S.vacuolatus)}}$	0.62$_{\text{(7d, L. minor)}}$		
SMTZ	>100$_{\text{(30 min)}}$	24.94$_{\text{(24h, S.vacuolatus)}}$	2.54$_{\text{(7d, L. minor)}}$		
SDMD	344.7$_{\text{(15 min)}}$ >100$_{\text{(30 min)}}$	19.52$_{\text{(24h, S.vacuolatus)}}$	1.277$_{\text{(7d, L. gibba)}}$ 1.74$_{\text{(7d, L. minor)}}$	174.4/158.8$_{\text{(48h/96h, D. magna)}}$ 215.9/506.3$_{\text{(48h/24h, D. magna)}}$ 111/311$_{\text{(48h/24h, M. macrocopa)}}$ 185.3/147.5$_{\text{(48h/96h, D. magna)}}$	>500$_{\text{(48h, O. latipes)}}$^{a} >500$_{\text{(96h, O. latipes)}}$^{a}
SMP	>100$_{\text{(30 min)}}$	3.82$_{\text{(24h, S.vacuolatus)}}$	1.51$_{\text{(7d, L. minor)}}$		
SCP	26.4$_{\text{(15 min)}}$ >50$_{\text{(30 min)}}$	32.25$_{\text{(24h, S.vacuolatus)}}$	2.33$_{\text{(7d, L. minor)}}$ 2.48$_{\text{(7d, L. minor)}}$	375.3/233.5$_{\text{(48h/96h, D. magna)}}$	589.3$_{\text{(48h, O. latipes)}}$^{a} 535.7$_{\text{(96h, O. latipes)}}$^{a}
SDM	>500$_{\text{(15 min)}}$ >500$_{\text{(5 min)}}$ >50$_{\text{(30 min)}}$	2.30$_{\text{(72h, P. subcapitata)}}$ 11.2$_{\text{(72h, C. vulgaris)}}$ 9.85$_{\text{(24h, S.vacuolatus)}}$	0.445$_{\text{(7d, L.gibba)}}$ 0.248$_{\text{(7d, L.gibba)}}$ 0.02$_{\text{(7d, L. minor)}}$	248.0/204.5$_{\text{(48h/96h, D. magna)}}$ 270/639.8$_{\text{(48h/24h, D. magna)}}$ 184/297$_{\text{(48h/24h, M. macrocopa)}}$	>100$_{\text{(48h, O. latipes)}}$^{a} >100$_{\text{(96h O. latipes)}}$^{a}
SQO ^{b}		0.25$_{\text{(96h, P. subcapitata)}}$ 0.45$_{\text{(96h, S. dimorphus)}}$ 2.83$_{\text{(96h, S. leopoliensis)}}$	13.55$_{\text{(7d, L.gibba)}}$ 2.33$_{\text{(7d, L. minor)}}$	3.47$_{\text{(48h, D. magna)}}$	

* green algae: *Pseudokirchneriella subcapitata* (previously *Scenedesmus capricornutum*), *Scenedesmus dimorphus*, *Chlorella vulgaris*; cyanobacteria *Synechococcus leopoliensis*, *Microcystis aeruginosa*; diatom *Cyclotella meneghiniana*;

** duckweed *Lemna gibba*, *Lemna minor*, carrot *Daucus carota*;

*** crustacean: *Moina macrocopa*, *Clathrina dubia*, *Thamnocephalus platyurus*, *Daphnia magna*; rotifer: *Brachionus calyciflorus*;

**** fish: *Oryzias latipes*, rainbow trout *Onchorhynchus mykiss*;

Table 2. Summary of the ecotoxicological risk (described by EC_{50} or $LC_{50}{}^{a}$ in mg L^{-1}) estimated for different sulfonamides (data obtained from [16,19-20,22-33]); ^{b} sulfaquinoxaline

This demonstrates the lack of data relating to the long-term exposure of non-target organisms, and especially how continuous exposure for several generations may affect a whole population. Moreover, as these compounds occur in natural media not as a single, isolated drug but usually together with other compounds of the same family or the same type, accumulated concentrations or synergistic-antagonistic effects can be also observed. The simultaneous presence of several pharmaceuticals in the environment may result in a higher level of toxicity towards non-target organisms than that predicted for individual active substances.

Therefore, the main aim of this chapter was to review the existing knowledge on the chronic and mixture toxicity of the residues of sulfonamides in the environment, since it has not been done yet. This will be achieved by: (1) presenting current approaches for Environmental Risk Assessment (ERA) for pharmaceuticals with respect to the evaluation of chronic and mixture toxicity of these compounds; (2) introducing the reader to basic concepts of chemical mixture toxicology; and finally (3) by discussing detailed available information on chronic and mixture toxicity of the residues of sulfonamides in the environment.

2. Environmental risk assessment of pharmaceuticals vs. chronic and mixture toxicity of pharmaceuticals

The approaches currently being used to assess the potential environmental effects of human and veterinary drugs in the U.S. and in the European Union are in some respects dissimilar [34-39]. The Environmental Risk Assessment (ERA) process usually starts with an initial exposure assessment (Phase I). But with some exceptions, a fate and effects analysis (Phase II) is only required when exposure-based threshold values, the so-called action limits, are exceeded in different environmental compartments. Thus risk assessment, described by Risk Quotient (RQ), is performed by the calculation the ratio of the predicted (or measured) environmental concentration (PEC or MEC respectively) and predicted biological non-effective concentrations (PNEC) on non-target organisms. If RQ is less than one it indicates that no further testing is recommended. Calculations of environmental concentrations rely e.g. on information on treatment dosage and intensity along with default values for standard husbandry practices, and are based on a total residue approach reflecting worst-case assumptions. For example, the recently introduced European guidance on assessing the risks of human drugs excludes the testing of pharmaceuticals whose $PEC_{surface\ water}$ is below an action limit of 0.01 $\mu g\ L^{-1}$; in the U.S. this threshold value is 0.1 $\mu g\ L^{-1}$. Moreover, there are two different action limits for veterinary pharmaceuticals, one each for the terrestrial and the aquatic compartments. No fate and effect analysis is required for veterinary pharmaceuticals used to treat animals if the PEC_{soil} is < 100 $\mu g\ kg^{-1}$ dry weight of soil. However, a Phase II assessment is not required for veterinary medicines used in an aquaculture facility if the estimated concentration of the compound is < 1 $\mu g\ L^{-1}$ [40-41]. If the $PEC_{surface\ water}$ of a pharmaceutical is above the action limit, effects on algae, crustaceans and fish are investigated. However, if PEC_{soil} is higher than the action limit, then Phase II, divided into two parts: Tier A, in which the possible fate of the pharmaceutical or its metabolites and its effects on earthworms (mortality) and plants (germination and growth) as well as the effects of the test compound on the rate of nitrate mineralization in soil are determined; and Tier B in which only effect studies are recommended for affected taxonomic levels [34-39].

The main problem associated with this approach is the fact that the no actual sales figures or measured environmental concentrations are at hand when a risk assessment is conducted. Therefore, only crude PEC calculations are performed [42]. Moreover, the (eco)toxicity tests included in Phase II focus on acute toxicity of only single compounds. Chronic and mixture toxicity is not obligatory. As the risk of an acute toxic effect from pharmaceuticals in the environment is unlikely and organisms in the environment are exposed to mixtures of pharmaceuticals, such limited focus results in important uncertainties. Additionally, same drugs (like sulfonamides) are used to treat both humans and animals. Although the exposures may differ, their potential effects on non-target organisms will be the same, and so the effect-testing approaches should be similar. For these reasons, many scientists have already pointed out the need for more reliable PEC and PNEC calculations for more realistic ERA of pharmaceutical [40-42].

3. Basic concepts of chemical mixture toxicology

To predict the toxicity of mixtures, ecotoxicologists use concepts originally developed by pharmacologists in the first half of the 20th century [43-48]. Since more than 20 years, they have been trying to elucidate the problem of risk assessment for complex mixtures of various substances. As a result a lot of excellent studies have been performed in this topic [49-51]. One of the main interests of scientists in the field of combination toxicology is to find out whether the toxicity of a mixture is different from the sum of the toxicities of the single compounds; in other words, will the toxic effect of a mixture be determined by additivity of dose or effect or by supra-additivity (synergism - an effect stronger than expected on the basis of additivity) or by infra-additivity (antagonism - an effect lower than the sum of the toxicities of the single compounds) The toxic effect of a mixture appears to be highly dependent on the dose (exposure level), the mechanism of action, and the target (receptor) of each of the mixture constituents. Thus, information on these aspects is a prerequisite for predicting the toxic effect of a mixture [46-47, 52].

Generally, three basic concepts for the description of the toxicological action of constituents of a mixture have been defined by Bliss and are still valid half a century later: (1) simple similar action (concentration addition, CA), (2) simple dissimilar action (independent action, IA), (3) interactions (synergism, potentiation, antagonism) [45].

Concentration addition (CA), also known as 'simple joint action', is based on the idea of a similar action of single compounds, whereas interpretations of this term can differ considerably. From mechanistic point of view, similar action means in a strict sense that single substance should show the same specific interaction with a molecular target site in the observed organisms. This is a nonintereactive process, which means that the chemicals in the mixture do not affect the toxicity of one another. Each of the chemicals in the mixture contributes to the toxicity of the mixture in proportion to its dose, expressed as the percentage of the dose of that chemical alone that would be required to obtain the given effect of the mixture. All chemicals of concern in a mixture act in the same way, by the same mechanisms, and differ only in their potencies [46-47, 52].

It has been shown that the concept of concentration addition is also applicable to nonreactive, nonionized organic chemicals, which show no specific mode of action but whose toxicity toward aquatic species is governed be hydrophobicity. The mode of action of such compounds is called narcosis or baseline toxicity [53-54]. The potency of a chemical to induce narcosis is entirely dependent on its hydrophobicity, generally expressed by its octanol-water partition coefficient $\log K_{ow}$. As a result, in the absence of any specific mechanism of toxicity, a chemical will, within certain boundaries, always be as toxic as its $\log K_{ow}$ indicates. Mathematically, the concept of concentration addition for a mixture of n substances is described by [48]:

$$\sum_{i=1}^{n} \frac{C_i}{ECx_i} = 1 \qquad (1)$$

where c_i represents the individual concentrations of the single substances present in a mixture with a total effect of $x\%$, and ECx_i are those concentrations of the single substances that would alone cause the same effect x as observed for the mixture. According to Eq. (1), the effect of the mixture remains constant when one component is replaced by an equal fraction of an equally effective concentration of another. As an important point, concentration addition means that substances applied at less than their individual "no observable effect concentrations" (NOECs) can nevertheless contribute to the total mixture effect [46-47].

The alternative concept of independent action (IA), also known as 'independent joint action' was already formulated by Bliss [45]. IA is when toxicants act independently and have different modes of toxic action [43, 46-47]. In this case the agents of a mixture do not affect each other's toxic effect. As a result of such a dissimiliar action, the relative effect of one of the toxicants in a mixture should remain unchanged in the presence of another one. For binary mixture the combination effect can be calculated by the equitation [46]:

$$E(c_{mix}) = 1 - [(1 - E(c_1))(1 - E(c_2))] \qquad (2)$$

In which $E(c_1)$, $E(c_2)$ are the effect of single substances and $E(c_{mix})$ is the total effect of the mixture. Following this equitation, a substance applied in a concentration below its individual NOEC will not contribute to the total effect of the mixture, i.e. there will be no mixture toxicity if the concentrations of all used single substances are below their NOEC [45-47, 52].

Additionally, compounds may interact with one another, modifying the magnitude and sometimes the nature of the toxic effect. This modification may make the composite effect stronger or weaker. An interaction might occur in the toxicokinetic phase (processes of uptake, distribution, metabolism, and excretion) or in the toxicodynamic phase (effects of chemicals on the receptor, cellular target, or organ).These include terms such as synergism and potentiation (i.e., resulting in a more than additive effect), or antagonism (i.e., resulting in a less than additive effect) [52]. It must be highlighted that at given concentrations of the single compounds in a mixture the combination effect will in general be higher if the substances follow the concept of concentration addition. Thus, misleadingly the different concepts were sometimes brought in correlation to the term synergism and antagonism. But synergism or antagonisms between the used substances and their effects can occur independently of a similar or dissimilar mode of action [46].

For these reasons, prediction of the effect of a mixture based on the knowledge of each of the constituents requires detailed information on the composition of the mixture, exposure level, mechanism of action, and receptor of the individual compounds. However, often such information is not or is only partially available and additional studies are needed. In addition to considering which of these concepts should be used to evaluate combined toxic effects, the design of the study is important in quantifying the combined effects. Most of such studies are based on a comparison of observed values with those predicted by a reference mode (IA or CA). An important aspect of toxicity studies of mixtures is the impracticability of 'complete' testing. If all combinations are to be studied at different dose levels, an increasing number of chemicals in a mixture results in an exponential increase in number of test groups: to test all possible combinations (in a complete experimental design) at only one dose level of each chemical in a mixture consisting of 4 or 6 chemicals, 16 (2^4–1) or 64 (2^6–1) test groups, respectively, would be required. Such *in vivo* studies are time consuming and impossible from a practical and economical point of view. To reduce the number of test groups without losing too much information about possible interactions between chemicals, several test scenario's (statistical designs) have been proposed [52]. The study design largely depends on the number of compounds of a mixture and on the question whether it is desirable to assess possible existing interactions between chemicals in a mixture [52]:

- One approach is to test the toxicity of the mixture without assessing the type of interactions. This is the simplest way to study effects of mixtures by comparing the effect of a mixture with the effects of all its constituents at comparable concentrations and duration of exposure at one dose level without testing all possible combinations of two or more chemicals. This approach requires a minimum number of experimental groups (n + 1, the number of compounds in a mixture plus the mixture itself). If there are no dose-effect curves of each of the single compounds it is impossible to describe the effect of the mixture in terms of synergism, potentiation, antagonism, etc. This strategy would be of interest for a first screening of adverse effects of a mixture.

- The second approach is based on assessment of interactive effects between two or three compounds which can be identified by physiologically based toxicokinetic modeling, isobolographic or dose-effect surface analysis, or comparison of dose-effect curves. However, interactive effects of compounds in mixtures with more than three compounds can be best ascertained with the help of statistical designs such as (fractionated) factorial designs, ray designs or dose-effect surface analysis. Here we would like to described shortly only the isobole methods as so far they are mainly used in the studies concerning the determination of pharmaceutical mixture toxicity [25-26,46-47].

An isobole, originally developed by Loewe and Muischnek [44], is a contour line that represents equi-effective quantities of two agents or their mixtures [52]. The theoretical line of additivity is the straight line connecting the individual doses of each of the single agents that produce the fixed effect alone. The method requires a number of mixtures to be tested and is used for a graphical representation to find out if mixtures of two compounds behave in a dose-additive manner and subsequently can be regarded as chemicals with a similar mode of action. When all equi-effect concentrations are connected by a downward concave line, the effect of the combinations is antagonistic, and a concave upward curve indicates synergism. The use of the isobole procedure to evaluate the effects of binary mixtures is widely used, but is very laborious and requires large data sets in order to produce sufficiently reliable results [52].

4. State of the knowledge concerning mixture and chronic toxicity of the residues of sulfonamides in the environment

4.1. What do we know about the long-term effects of the presence of the residues of sulfonamides in the environment?

Chronic toxicity tests are studies in which organisms are exposed to different concentrations of a chemical and observed over a long period, or a substantial part of their lifespan. In contrast to acute toxicity tests, which often use mortality as the only measured effect, chronic tests usually include additional measures of effect such as growth rates, reproduction or changes in organism behavior [55-56]. Therefore, the standard acute toxicity tests do not seem appropriate for risk assessment of pharmaceuticals, because of the nature of these compounds. The use of chronic tests over the life-cycle of organisms for different trophic levels could be more appropriate [57]. However, there is still an ongoing debate between ecotoxicologists over the determination which tests should be considered to be chronic or acute (based on their duration). This applies not only to aquatic animal testing with invertebrates and fish, but also to standard 96-h algal and 7-d higher plant test methods.

Molander et al. [19] reviewed the data published in the *Wikipharma* database – a freely available, interactive and comprehensive database on the environmental effects of pharmaceuticals that provides an overview of effects caused by these compounds on non-target organisms identified in acute, sub-chronic and chronic ecotoxicity tests. Looking at the data set as a whole, they concluded that crustaceans like *Daphnia magna* and *Ceriodaphnia dubia* were the species most commonly used (29% of all tests performed); this is hardly surprising since they are abundant and widespread, easy to keep in the laboratory, and sensitive towards a broad range of environmental contaminants. Less commonly, such tests were performed on marine bacteria *Vibrio fischeri* (12%), algae *Pseudokirchneriella subcapitata* (9.5%) and fish *Poeciliopsis lucida* (9%) and *Oncorhynchus mykiss* (8%) [19]. They have also estimated that acute tests based on microorganisms (exposure time ≤ 30 min), algae (exposure time ≤ 72 h), invertebrates (exposure time ≤ 48 h) and vertebrates (exposure time ≤ 96 h) constitute 55% of all the data compiled [19]. This information was corroborated by Santos et al. [20], who estimated that acute effects in organisms belonging to different trophic levels predominate over chronic ones in more than 60% of all the tests performed. This also concerns the available information on the ecotoxicity of sulfonamides (see Table 2).

Looking at the available acute toxicity data, it can be concluded that SAs are practically non-toxic to most microorganisms tested including selected strains of bacteria, such as *Vibrio fischeri* and *Pseudomonas aeruginosa*. However, data as are available from acute tests on the potential effects of SAs in the environment appear to indicate a possible negative impact on different ecosystems and imply a threat to public health. The most sensitive assays for the presence of SAs are bioindicators containing chlorophyll (algea and duckweed) [3, 22-23]. A highly toxic effect of SMX on duckweed (*Lemna gibba*) was observed. This was also supported by the results of one of our studies, where we evaluated the ecotoxicity potential of twelve sulfonamides (sulfaguanidine, sulfadiazine, sulfathiazole, sulfamerazine, sulfamethiazole, sulfachloropyridazine, sulfamethoxypyridazine, sulfamethoxazole, sulfisoxazole, sulfadimethoxine, sulfapyridine, sulfadimidine) to enzymes (acetylcholinesterase and glutathione reductase), luminescent marine bacteria (*Vibrio fischeri*), soil bacteria (*Arthrobacter globiformis*), limnic unicellular green algae (*Scenedesmus vacuolatus*) and duckweed (*Lemna minor*). We found that SAs were not only toxic towards green algae (EC_{50} = 1.54 – 32.25 mg/L) but were even more strongly so towards duckweed (EC_{50} = 0.02 – 4.89 mg/L) than atrazine, a herbicide (EC_{50} = 2.59 mg/L) [33]. This indicates that even low concentrations of SAs may significantly affect the growth and development of plants.

However, data relating to the long-term exposure of non-target organisms, and especially how continuous exposure for several generations may affect a whole population is very limited. Most chronic toxicity data for sulfonamides, is available for invertebrates, probably because these are the briefest and therefore least expensive chronic toxicity tests to run. Available chronic toxicity data for sulfonamides is summarized in Table 3 and discussed below.

The major concern over the effects of all antimicrobials (including sulfonamides) on microbial assemblages is the development of antimicrobial resistance and the effect of this on public health. Recently, Baran et al. [3] has reviewed the papers concerning the influence of presence of SAs in the environment to antimicrobial resistance. They concluded that SAs in the environment increase the antimicrobial resistance of microorganisms and the number of bacterial strains resistant to SAs increases systematically in recent years. Resistant bacterial species commonly carried single genes, but in recent years, an increased number of pathogens that possess three SAs-resistant genes have been observed. Moreover, they have also highlighted that these drugs have shown the highest drug resistance, almost twice as high as tetracyclines and many times higher than other antibiotics. Most often, bacterial resistance to SAs has been described in *Escherichia coli, Salmonella enterica* and *Shigella spp.* from the manure of farm animals, from meat and from wastewater [3]. The implications of antimicrobial resistance for aquatic ecosystem structure and function remain unknown, but the human health implications of widespread resistance are of clear concern [55].

Additionaly, Heuer and Smalla [58] investigated the effects of pig manure and sulfadiazine on bacterial communities in soil microcosms using two soil types. In both soils, manure and sulfadiazine positively affected the quotients of total and sulfadiazine-resistant culturable bacteria after two months. The results suggest that manure from treated pigs enhances spread of antibiotic resistances in soil bacterial communities. Monteiro and Boxall [59] have recently examined the indirect effects of sulfamethoxazole on the degradation of a range of human medicines in soils. It was observed that the addition of SMX significantly reduce the rate of degradation of human non-steroidal anti-inflammatory drugs, naproxen. This observation may have serious implications for the risks of other compounds that are applied to the soil environment such as pesticides.

Substance name	Type of organism	Acute toxicity	Chronic toxicity	Ref.
SDZ	*Daphnia magna*	$EC_{50,\,48h}$ = 221 mg L^{-1} (166 – 568 mg L^{-1})	$EC_{50,\,21d}$ = 13.7 mg L^{-1} (12.2 – 15.3 mg L^{-1})	[60]
SMX	*Brachionus calyciflorus*	$EC_{50,\,24h}$ = 26.27 mg L^{-1} (16.32 – 42.28 mg L^{-1})	$EC_{50,\,48h}$ = 9.63 mg L^{-1} (7.00 – 13.25 mg L^{-1})	[29]
	Clathrina dubia	$EC_{50,\,48h}$ = 15.51 mg L^{-1} (12.97 – 18.55 mg L^{-1}	$EC_{50,\,7d}$ = 0.21 mg L^{-1} (0.14 – 0.39 mg L^{-1})	
		$-logEC_{50,\,15\,min}$	$-logEC_{50,\,24h}$	
SDMD		3.12 (± 0.04) M	4.08 (± 0.06) M	
SPY		2.92 (± 0.05) M	3.84 (± 0.04) M	
SMX		3.32 (± 0.04) M	4.45 (± 0.05) M	
SDZ	*Photobacterium phosphoreum*	3.32 (± 0.02) M	4.50 (± 0.06) M	[61]
SSX		3.81 (± 0.02) M	4.43 (± 0.03) M	
SMM[a]		3.67 (± 0.03) M	5.05 (± 0.05) M	
SCP		4.30 (± 0.04) M	4.78 (± 0.04) M	
SQO		$EC_{50,\,48h}$ = 131 mg L^{-1} (119 – 143 mg L^{-1})	$EC_{50,\,21d}$ = 3.466 mg L^{-1} (2.642 – 4.469 mg L^{-1})	
SGD	*Daphnia magna*	$EC_{50,\,48h}$ = 3.86 mg L^{-1} (3.19 – 5.08 mg L^{-1})	$EC_{50,\,21d}$ = 0.869 mg L^{-1} (0.630 – 1.097 mg L^{-1})	[25-26]
SDMD		$EC_{50,\,48h}$ = 202 mg L^{-1} (179 – 223 mg L^{-1})	$EC_{50,\,21d}$ = 4.25 mg L^{-1} (3.84 – 4.62 mg L^{-1})	
SDMD	*Daphnia magna*	$EC_{50,\,48h}$ = 215.9 mg L^{-1} (169.6 – 274.9 mg L^{-1})	$EC_{50,\,21d}$ no effect up to 30 mg L^{-1}	
STZ		$EC_{50,\,48h}$ = 616.9 mg L^{-1} (291.7 – 1303.6 mg L^{-1})	LOEC = 35 mg L^{-1}	[16]
SDMD	*Moina macrocopa*	$EC_{50,\,48h}$ = 110.7 mg L^{-1} (89.5 – 136.9 mg L^{-1})	$EC_{50,\,8d}$ no effect up to 35 mg L^{-1}	
STZ		$EC_{50,\,48h}$ = 391.1 mg L^{-1} (341.9 – 440.3 mg L^{-1})	no effect up to 30 mg L^{-1}	

[a]Sulfamonometoxine

Table 3. An overview of the available information on the chronic toxicity of sulfonamides to different organisms

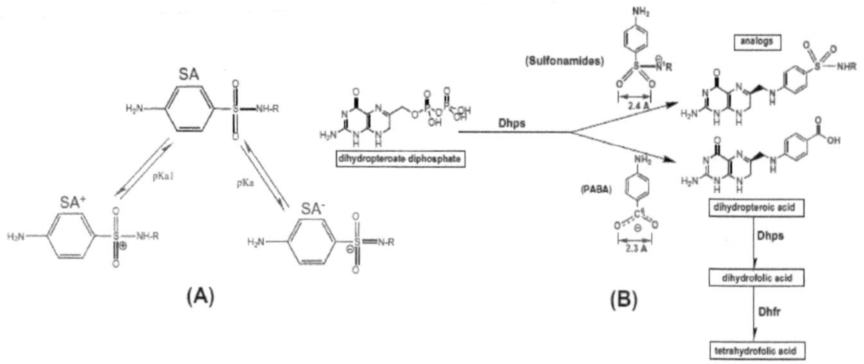

Figure 1. (A) Scheme of SA ionization in equilibrium, (B) Mechanism for synergistic effect between SA and TMP in bacterium (adopted from Reference [61])

Only few studies have also explored effects of SAs on aquatic microbes. It must be highlighted that it was already proved that the effects of antibiotics like SAs on bacteria should not be determined using acute tests. These compounds possess specific mode of action and impacts frequently became evident upon extending the incubation period. Most of the toxicity data available for *Vibrio fischeri* using short exposure times (between 5 and 30 min) rather than a 24 h exposure show that SAs have a low toxic potential in this respect, because these compounds interfere only slightly with biosynthetic pathways. Toxicity tests with bacteria have shown that chronic exposure to antibiotics is crucial rather than acute [14,50, 62-63]. This is also supported by the results of [30,61,64]. The toxicity of sulfadimidine (sulfamethazine) in standard 15 min acute test with this luminescent bacteria obtained EC_{50} was 344.7 mg L^{-1} [30] but in 18 h test its toxicity was in the range of 3.68 – 4.57 mg L^{-1} depending on the type of strain of these marine bacteria [64]. Also Zou et al. [61] determine the chronic (24 h exposure) and acute (15 min exposure) toxicity to *Photobacterium phosphoreum* for seven SAs. These experiments revealed that sulfachloropyridazine (SCP) was more toxic than other SAs, whereas sulfapyridine (SPY) was relatively less toxic than other SAs (see Table 3). The order of acute toxicity was as follows: SCP > SSX > SMX > SMM > SDZ > SDMD > SPY. However, the order of chronic toxicity was different: SMM > SCP > SDZ > SMX > SSX > SDMD > SPY. Clearly, different order of toxicity between the acute and chronic exposure indicated a different toxicity mechanism (see Fig. 1). It has been reported that the acute toxic effects of pollutants to *P. phosphoreum* are caused by interfering LUC-catalyzed bioluminescent reaction and therefore LUC was found to be the receptor protein for the antibiotics. In contrast, the receptor for the antibiotics in the chronic toxicity test was dihydropterinic acid synthetase (DHPS).

Studies conducted on the toxicity mechanism of single SAa indicated that the pKa played a vital role in the toxic effect of SAs or their antibacterial activity [32]. Because LUC (Lucyferase) is an endoenzyme, and SAs have to be transported into the cell before bind with LUC, it was clear that the antibiotic toxicity included both LUC-binding and a toxic transportation effect (which can be described using pKa). pKa is a decisive factor in transporting SAs into the cell. Three species (neutral, cationic and anionic) of SAs depend on the pKa and surrounding pH values. The neutral species have higher cell membrane permeability than anionic species. Therefore, pKa was the key parameter of sulfonamides toxic effects. Some similarity in acute and chronic toxicity mechanisms was observed. However, in conclusion the distinct receptor proteins of SAs in acute toxicity and chronic toxicity led to the different toxicity mechanisms of single antibiotics [61]. A comparison of the results of short and long term bioassays with *Vibrio fischeri* demonstrates the risk of underestimating the severe effects of substances with delayed toxicity in acute tests.

Similar conclusions can be obtained if only acute toxicity of SAs to invertebrates is taken into consideration. Detailed information is presented in Table 3. No acute effects on *D. magna* were observed in the investigation of Wollenberger et al. [60]. However, reproductive effects (EC_{50}) were observed for sulfadiazine in the range of 5 to 50 mg L^{-1}. This drug caused mortality in the parent generation during the long-term (3 weeks) exposure. Such results suggest that crustacean reproduction test should be included in the test strategy [60]. Similar correlation was also found by Isidori et al. [29] who investigated the acute and chronic toxicity of SMX to *B. calciflorus* (24 h and 48 h exposure) and *C. dubia* (48 h and 7 d exposure time). As expected chronic tests showed higher toxicity that acute tests. Also Park and Choi [16] evaluated the acute and chronic aquatic toxicities of four SAa using standard tests with *D. magna* and *M. macrocopa*. The results from the chronic toxicity tests in this study showed that sensitivity of *M. macrocopa* was similar to that of *D. magna*. However, the exposure duration for *M. macrocopa* was only 8 days whereas for D. magna was 21 days. *Moina* shares many characterisitcs with *D. magna* (e.g. large population densities, high population growth rates, short generation time, and easiness of culture) and is often preferred for hazard evaluation because of its relatively short life span and wide geographical distribution. However, Park and Choi found no significant effects on reproduction of *D. magna* at concentrations of SMZ up to 30 mg L-1. In contrast De Liguoro et al. [25-26] observed strong inhibitory effect of SMZ on reproduction of *D. magna* (nearly 100% inhibition with SMZ at a concentration of 12.5 mg L-1). This could be explained by that fact that in the Park and Choi study, daphnids were fed daily not only with algae, but also with the EPA recommended YCT that contains yeast, a known good natural source of folate [16]. Eguchi et al. [28] have shown that SAs interfere with folate synthesis in green algae. Therefore, this supplement of folic acid may well have protected the reproduction of the test organisms by compensating for the deficiencies caused by SMZ. Generally, when testing antibacterials on *D. magna*, effects on the reproductive output occur at concentrations which are at least one order of magnitude below the acute toxic levels [60].

Unfortunately, there is no information about long-term effects of the residues of these compounds to higher plants and other aquatic as well as terrestrial organisms. Therefore, it seems to be necessary for researchers to study the chronic toxicity of antibiotic [46-47, 55] because of their widespread use and continuous emissions into the environment [14].

4.2. What do we know about mixture toxicity of the residues of sulfonamides in the environment?

As SAs occur in natural environment not as a single, isolated drug but usually together with other compounds of the same family or the same type, accumulated concentrations or synergistic-antagonistic effects need to be considered. Sulfonamides are widely used in combination therapy together with their potentiator (mostly trimethoprim, TMP) in human and veterinary medicine [1-3]; thus, the occurrence of TMP together with other antibiotics has been commonly detected [3].

Santos et al. [20] pointed out that, ecotoxicological data show that the effects of mixtures may differ from those of single compounds. For example, Cleuvers [46] showed that a mixture of diclofenac and ibuprofen exhibited a greater than predicted toxicity to *D. magna*, and that the addition of two more drugs increased the toxicity towards the test species even further. Available mixture toxicity data for sulfonamides is summarized in Table 4 and discussed below.

Substance name/ Mixture composition	Test scenario (statistical design)	Test description (organism, test duration)	Toxicity of single compounds ($EC_{50single}$)	Mixture toxicity ($EC_{50mixture}$)	Conclusions	Ref.
SDM	Evaluation of the toxicity of the mixture of selected compounds based on concentration addition concept.	*Selenastrum capricornutum*, according to OECD 201	2.30 mg L^{-1}		Synergistic growth inhibition between SAs and TMP or PMT and for SMX:AcSMX:TMP mixture.	[28]
SMX			1.50 mg L^{-1}			
SDZ			2.19 mg L^{-1}			
Trimethoprim (TMP)			80.8 mg L^{-1}			
Pyrimethanine (PMT)			5.06 mg L^{-1}			
AcSMX			>100mg L^{-1}			
AcSDM			>100mg L^{-1}			
AcSDZ			>100mg L^{-1}			
SMZ + TMP				0.275 mg L^{-1}		
SDA + TMP				0.465 mg L^{-1}		
SDM + PMT				2.36 mg L^{-1}		
SMX:AcSMX:TMP (20:105:3)				0.784 mg L^{-1}		
SDM:AcSDM:TMP (176:8:1)				2.17 mg L^{-1}		
SDZ:AcSDZ:TMP (42:24:1)				2.08 mg L^{-1}		
SDZ	Assessment of interactive effects between two compounds identified by	*Daphnia magna*, 48 h	212 mg L^{-1}		Antagonistic interaction for mixtures: SMZ + SDZ SMZ + SGD SMZ + SMR	[26]
SGD			3.86 mg L^{-1}			
SMR			277 mg L^{-1}			
SDM			202 mg L^{-1}			
SDMD (SMZ)			270 mg L^{-1}			

SQO			131 mg L^{-1}		SMT + SDM Complex interaction (synergism additivity and antagonism) for mixture of SMT + SQO	
TMP			149 mg L^{-1}			
binary mixtures of SMZ +6 compounds	isobologram method.				Simple additivity for SMZ + TMP	
SGD	Assessment of interactive effects between two compounds identified by isobologram method.	*Daphnia magna*, 21 d	0.896 mg L^{-1}		Additive (antagonistic) interaction between SQO and SGD.	[25]
SQO			3.466 mg L^{-1}			
SGD		*Pseudokirchneriella subcapitata*, 96 h	43.559 mg L^{-1}			
SQO			0.246 mg L^{-1}			
			Toxicity of single compound ($-logEC_{50, 15 min}$)	Toxicity of binary mixture SAs and TMP($-logEC_{50, mixture}$)		
SDMD	Evaluation of the toxicity of the mixture of selected compounds based on concentration addition concept	*Photobacterium phosphoreum*, 15 min and 24 h	3.12 (± 0.04) M	2.78 (± 0.02) M	Antagonistic interaction between SAs and TMP in acute toxicity test.	[61]
SPY			2.92 (± 0.05) M	2.89 (± 0.02) M		
SMX			3.32 (± 0.04) M	2.79 (± 0.01) M		
SDZ			3.32 (± 0.02) M	2.76 (± 0.03) M		
SSX			3.81 (± 0.02) M	2.79 (± 0.07) M		
SMM			3.67 (± 0.03) M	2.73 (± 0.02) M		
SCP			4.30 (± 0.04) M	3.00 (± 0.03) M		
TMP			3.22 (± 0.07) M			
			Toxicity of single compound ($-logEC_{50, 24h}$)	Toxicity of binary mixture SAs and TMP ($-logEC_{50, mixture}$)		
SDMD			4.08 (± 0.06) M	5.08 (± 0.05) M	Synergistic interaction between SAs and TMP in chronic toxicity test.	
SPY			3.84 (± 0.04) M	4.85 (± 0.07) M		
SMX			4.45 (± 0.05) M	5.50 (± 0.07) M		
SDZ			4.50 (± 0.06) M	5.42 (± 0.03) M		
SSX			4.43 (± 0.03) M	5.45 (± 0.03) M		
SMM			5.05 (± 0.05) M	6.01 (± 0.05) M		
SCP			4.78 (± 0.04) M	5.73 (± 0.05) M		
TMP			5.37 (± 0.02) M			

Table 4. An overview of the available information on the mixture toxicity of sulfonamides to different organisms

The toxicity of mixture of sulfonamides to non-target organisms was firstly reported by Brain et al. [22] and Eguchi et al. [28]. Brain et al. investigated the toxicity of the mixture of eight most commonly used pharmaceuticals belonging to different groups (atorvastatin, acetaminophen, caffeine, sulfamethoxazole, carbamazepine, levofloxacin, sertraline and trimethoprim) to the aquatic macrophytes *Lemna gibba* and *Myriophyllum sibircum*. Given the diversity in mode of action of these compounds, the toxicity of the mixture in the microcosms was likely via response addition. Generally, both species displayed similar sensitivity to the pharmaceutical mixture [22].

On the other hand, Eguchi et al. [28] found that a mixture of trimethoprim or pyrimethamine (pyrimethamine is often used as a substitute for trimethoprim), sulfamethoxazole and sulfadiazine significantly increased growth inhibition (synergistic effect of the mixture was observed) in the algae *S. capricornutum*. To investigate the synergistic influence of combined drugs on the growth of green algae, SAs and TMP or PMT (TMPs) were simultaneously added to *S. capricornutum* culture. In this experiment, the concentration of TMPs was fixed at the no observed effect concentration (NOEC) and the concentration of the SAs were altered. These combined drugs are frequently used in the veterinary field in many countries. Combination of SMZ and SDA with TMP rendered the growth inhibitory activity significantly increased in comparison with their individual activities (see Table 4 and Fig. 2(A)). On the contrary, combination of SDM with PYR did not show such an effect. Moreover, as SAs are thought to be partly metabolized to AcSAs in the bodies of animals, Eguchi et al. [28] have also tested the toxicity of the mixture of SAs their metabolites and TMP. Therefore the test of combined drugs was done by using the combinations corresponding AcSAs at a ratio according to the concentrations detected in the urine of pigs fed with SAs. The ratio was SMZ:AcSMZ:TMP = 20:105:3, SDM:AcSDM:PMT = 167:8:1 and SDA:AcSDA:TMP = 42:24:1. A similar synergistic effect to that described above was observed with combinations of SMZ, TMP, and AcSMZ (see Table 4). However, combination of SDM or SDA with their acetylate and PMT or TMP did not show a synergistic effect on growth in excretion ratio. A reason must be that the concentration of TMP used was not enough to express synergistic influence in combination with SDM or SDA. These results indicate that several combined drugs that show a synergistic effect *in vitro* may have an actual synergistic effect on algae in ecosystem although excretion ratio can vary in animal condition or other factors. The synergistic effect observed by the combination of SAs and TMPs in this study indicates that the simultaneous release of several antimicrobial agents may result in greater toxicity to microorganisms in the environment than the release of the same agents individually. Furthermore, the rate of growth inhibition by SAs by addition of folic acid was investigated in this study. It was observed that the growth inhibitory activity of the combination of SDA and TMP was significantly reduced by the addition of 20 ng/l of folic acid to the medium. Significantly, folic acid exhibited a similar effect when SDA was tested alone, but not when TMP was tested alone (see Fig. 2(B)) [28].

Both SAs and TMPs inhibit the folate synthesis pathway in bacteria, but their inhibition sites are different. SAs inhibit dihydropterinic acid synthetase (DHPS), thereby inhibiting the synthesis of folic acid. On the other hand, TMPs inhibits dihydrofolic acid reductase (DHFR), which converts folic acid to 7,8-dihydrofolic acid (7,8- DHF) and 5,6,7,8-tetrahydrofolic acid (5,6,7,8-THF), both active forms of folic acid suitable for utilization. Therefore, the synergistic effect of the combination of SAs and TMPs is likely to be due to the cumulative effect of their

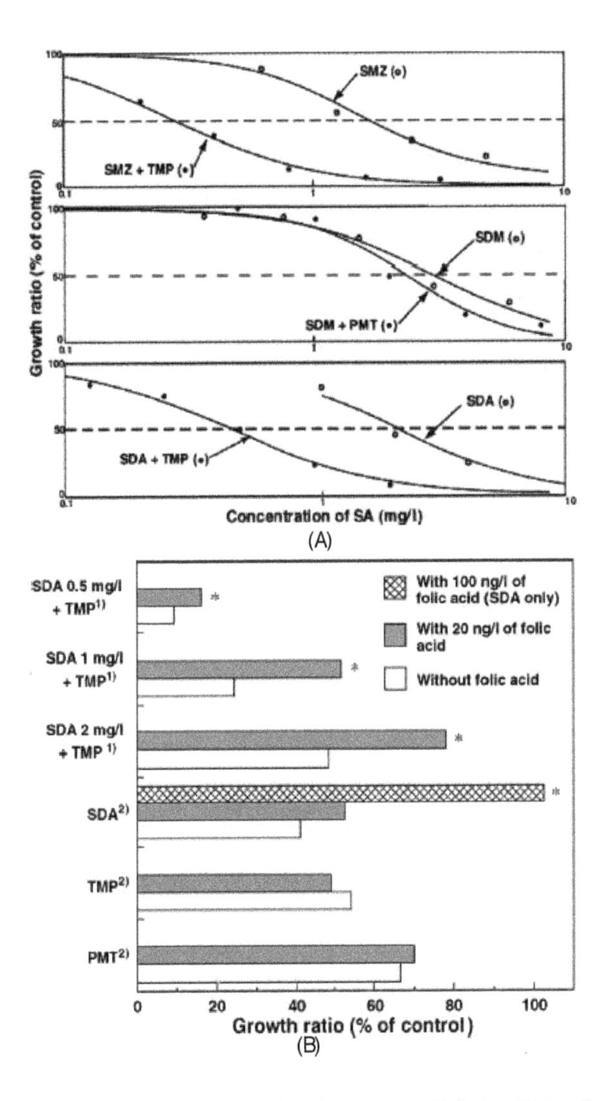

Figure 2. (A) The dose-response curve of SAs (SMT – sufamethzine, SDA – sulfadiazine, SDM – sulfadimethoxine) combined with TMP (trimethoprim) or PMT (pyrimethamine); **(B)** Recovery of growth inhibition be addition of folic acid (*observed siginifcant difference to negative control – without folic acid, [1] concentration of SDA (sulfadiazine) in combination, TMP was used at the NOEC, [2] used at the EC_{50} concentration) (adopted from References [28])

actions on two different sites in the folate biosynthesis pathway. Since SAs block the synthesis of folate, the growth inhibitory effect of this compound can be reversed by the addition of folate. In contrast, TMP blocks enzymes downstream of folate in the synthesis pathway, thus addition of folate will not reverse the growth-inhibiting effect of this compound. Since algea also have a similar folate synthesis pathway, the growth inhibitory effect of SAs on these organisms is likely to be the result of the same inhibitory mechanism. Therefore, algal cells could survive in the presence of SAs, but not TMP, when folic acid was added to the medium.

De Liguoro et al. [25-26] evaluated the acute mixture toxicity of combining sulfamethazine with TMP towards *D. magna* and effects of different mixtures of sulfaquinoloxine (SQO) and sulfaguanidine (SGD) on *D. magna* and *P. subcapitata* (see Table 4). The additive toxicity of these compounds was evaluated using the isobologram method. In Fig. 3A, the isoboles showing the different type of combination effects are presented. Taking into account confidence intervals SMZ showed infra-additivity when paired with SDZ, SGD, SMA or SDM. When SMZ was paired with SQO the interaction was more complex, as each type of combination effects (supra-additivity, additivity and infra-additivity) was observed at the three different combination ratios. Simple additivity was recorded when SMZ was combined with the sulfonamide-potentiator TMP (Fig. 3A). Tests with paired SQO and SGD were based on the individual EC_{50} (for *D. magna* see [25]). In each paired test, the concentration–response relationship was analyzed for three selected combination ratios equidistantly distributed on the additivity line. In Fig. 3B, the isoboles based on the effects of different mixtures of SQO and SGD on *D. magna* and *P. subcapitata* are depicted. Only in one test, where relatively low concentrations of SQO were combined with relatively high concentrations of SGD on *P. subcapitata*, the two paired compounds showed simple additivity. In all the other tests a less than additive (antagonistic) interaction was detected. In this study, binary tests confirmed the tendency of SAs mixtures to act less than additively. So, in general terms, it seems sufficiently precautionary to consider their environmental toxicity as additive. However, when combining SQO and SGD on P. *subcapitata*, the obtained asymmetric isobologram shows that the interaction is mixture-ratio dependent, a phenomenon already observed when mixtures of SQO and sulfamethazine were tested on *D. magna* [26].

Zou et al. [61] have recently highlighted that these results cannot represent the mixture toxicity between the SAs and TMP in an actual environment because non-target organisms (microlage and *D. magna*) were used in these studies. Bacterium is typically the target-organism of an antibiotic, and thus in their opinion, a bioassay with *Photobacterium phosphoreum* is a more reliable tool to determine the toxicity of various antibiotics. Moreover, most studies focus on the acute mixture toxicity. Therefore, in their study they have: determined not only the acute (15 min exposure) but also chronic (24 h exposure) toxicity *to P. phosphoreum* for single SA and their potentiator, and for their mixtures (SA with TMP); evaluated the differences between chronic and acute mixture toxicity; and revealed the difference between their toxicity mechanisms by using QSAR models. A comparison of chronic vs. acute mixture toxicity revealed the presence of an interesting phenomenon, that is, that the joint effects vary with the duration of exposure; the acute mixture toxicity was antagonistic, whereas the chronic mixture toxicity was synergistic. Based on the approach of QSARs and molecular docking, this phenomenon was proved to be caused by the presence of two points of dissimilarity between the acute and chronic mixture toxicity mechanism: (1) the receptor protein of SAs in acute toxicity was LUC, while in chronic toxicity it was DHPS, and (2) there is a difference between actual concentration of binding-LUC in acute toxicity and individual binding-DHPS in chronic toxicity (see Fig. 1). The existence of these differences poses a challenge for the assessment of routine combinations in medicine, risk assessment, and mixture pollutant control, in which, previously, only

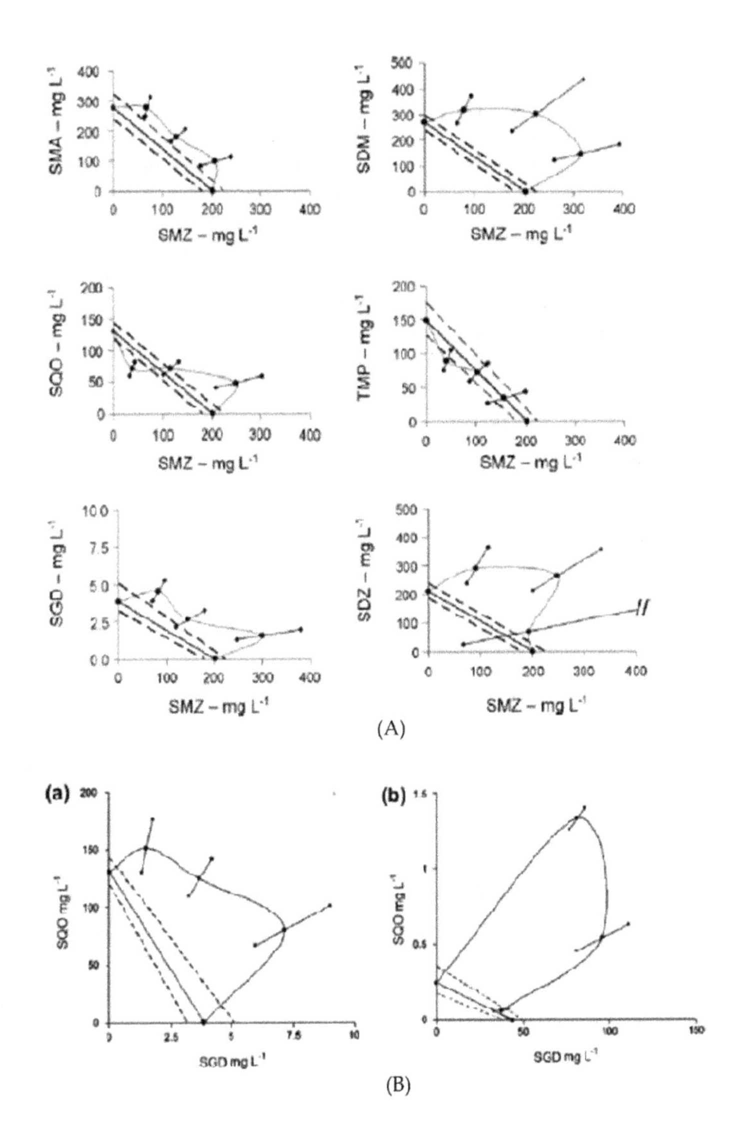

Figure 3. (A) *D. magna* immobilization test: 48 h EC_{50} isobolograms of SMZ paired with other SAs and TMP at three selected combination ratios; **(B)** isobolograms of paired SQO and SGD *in D. magna* immobilization test (adopted from References [25-26])

a synergistic effect has been observed between SA and their potentiator. The toxicity effect of mixtures is associated with the transportation of toxic effects of individual chemicals into cells, the interaction toxic effects of individual chemical-binding-receptor proteins. According to acute toxicity mechanism of single antibiotics it can be concluded that the transportation toxic effect is highly related to pKa values and the interaction toxic effect can be described by LUC-SAs binding or LUC-TMP binding. The synergistic effect between SAs with TMP has also been observed in the field of medicine and proved to be caused by the blocking of synthesis of folic acid. First SAs inhibit DHPS, which catalyzes the formation of dihydropteroic acid. Then TMP inhibits DHFR, which catalyzes the formation of tetrahydrofolic acid from dihydrofolic acid. In an acute mixture toxicity there were more SAs-binding-LUC and TMP-binding-LUC. However, in chronic mixture toxicity, the concentration of SAs-binding-DHPS was less compere to TMP-binding-DHFR. It can therefore be concluded that the dissimilarities in the concentrations of individual chemical-binding receptor proteins also lead to the different joint effect (SA with TMP) in acute and chronic mixture toxicity [61].

These examples highlight the fact that the simultaneous presence of several pharmaceuticals in the environment may result in a higher level of toxicity towards non-target organisms than that predicted for individual active substances. More ecotoxicological studies should therefore be done to evaluate the impact of different mixtures of pharmaceuticals in non-target organisms.

5. Conclusions

The reason for concern regarding risks of mixtures is obvious. Man is always exposed to more than one chemical at a time. This dictates the necessity of exposure assessment, hazard identification, and risk assessment of chemical mixtures. However, for most chemical mixtures data on exposure and toxicity are fragmentary, and roughly over 95% of the resources in toxicology is still devoted to studies of single chemicals. Moreover, organisms are typically exposed to mixtures of chemicals over long periods of time; thus, chronic mixture toxicity analysis is the best way to perform risk assessment in regards to organisms.

However, testing of all kinds of (complex) mixtures of chemicals existing in the real world or of all possible combinations of chemicals of a simple (defined) mixture at different dose levels is virtually impossible. Moreover, even if toxicity data on individual compounds are available, we are still facing the immense problem of extrapolation of findings obtained at relatively high exposure concentration in laboratory animals to man being exposed to (much) lower concentrations.

As stated by several authors, it is essential to investigate if mixtures of pharmaceuticals interact, leading to a larger effect in the environment than would be predicted when each compound is considered individually. Mixtures with antibiotics in the environment may be very complex (e.g. wastewater effluent) but they also may be simple. Although the latter may be more easily studied experimentally, in both cases the identification and quantitative description of synergism caused by specific substances is crucial.

Over past 10 years there has been increasing interest in the impacts of SAs and other veterinary medicines in the environment and there is now a much better understanding about their environmental fate and their impacts on aquatic and terrestrial organisms. However, there are still a number of uncertainties that require addressing before there can be a full understanding of the environmental risks of these compounds. Areas requiring further research are presented below.

- The assessment of the potential impacts of those SAs for which ecotoxicity data is lacking but are seen to regularly occur in the environment.
- More information about the ecotoxicity of these compounds to soil organisms should be provided. This regards to acute, chronic and single/mixture toxicity of most of the veterinary pharmaceuticals.
- Information on the potential environmental effects of parent compounds (drugs) as well as metabolites and transformation products. This includes the single and joint effects evaluation.
- Further research is required on the mixture toxicity of SAs in combination with other medicines and non-medicinal substances.
- The possible indirect effects of SAs should be identified.
- Data from acute and chronic ecotoxicity tests on species belonging to different trophic levels such as bacteria, algea, crustaceans and fish among others, is relevant to illustrate the several adverse effects that environmental exposure to measured concentrations of these contaminants can have. The principal toxicological endpoints/studies that are described are growth, survival, reproduction and immobilization of species, comparatively to trangenerational and population level studies that are still sparse. In the near future, the evaluation of chronic toxicity effects should be set out as a priority for the scientific community since simultaneous exposure to pharmaceuticals, metabolites and transformation products of several therapeutic classes are unknown and whose probable effects on subsequent generations should be assumed.

Abbreviations

Abbreviation	Full name
CA	Concentration Addition
DHFR	Dihydrofolic Acid Reductase
DHPS	Dihydropterinic Acid Synthetase
EPA	Environmental Protection Agency
ERA	Environmental Risk Assessment
IA	Independent Action
LUC	Lucyferase
MEC	Measured Environmental Concentration
NOEC	No Observable Effect Concentrations
PABA	*p*-aminobenzoic acid

Abbreviation	Full name
PEC	Predicted Environmental Concentration
PMT	Pyrimethamine
PNEC	Predicted Non-Effective Concentrations
RQ	Risk Quotient
SAs	Sulfonamides
SCP	Sulfachloropyridazine
SDM	Sulfadimethoxine
SDMD (SMZ)	Sulfadimidine (Sulfamethazine)
SDZ	Sulfadiazine
SGD	Sulfaguanidine
SMP	Sulfamethoxypyridazine
SMR	Sulfamerazine
SMTZ	Sulfamethiazole
SMX	Sulfamethoxazole
SPY	Sulfapyridine
SQO	Sulfaquinoxaline
SSX	Sulfisoxazole
STZ	Sulfathiazole
TMP	Trimethoprim

Table 5. List of abbreviations used in the text

Acknowledgements

Financial support was provided by the Polish National Science Centre under grant DEC-2011/03/B/NZ8/03009 "Determining the potential effects of pharmaceuticals in the environment: an ecotoxicity evaluation of selected veterinary drugs and their mixtures" (2012-2015).

Author details

Anna Białk-Bielińska, Jolanta Kumirska and Piotr Stepnowski

Department of Environmental Analysis, Faculty of Chemistry, University of Gdańsk, Gdańsk, Poland

References

[1] Sukul P., Spiteller M. Sulfonamides in the environment as veterinary drugs. Reviews of Environmental Contamination and Toxicology 2006; 187 67-101.

[2] García-Galán MJ., Díaz-Cruz MS., Barceló D. Identification and determination of metabolites and degradation products of sulfonamide antibiotics. Trends in Analytical Chemistry 2008; 27 1008-1022.

[3] Baran W., Adamek E., Ziemiańska J., Sobczak A. Effects of the presence of sulfonamides in the environment and their influence on human health. Journal of Hazardous Materials 2011; 196 1-15.

[4] Briuce PY. Organic Chemistry. Prentice-Hall, Inc., Upper Saddle River; 1995.

[5] Bell PH., Roblin RO. Studies in chemotherapy. Journal of the American Chemical Society1942;64 2905-2917.

[6] Şanli S., Altun Y., Şanli N., Alsancak G., Baltran JL. Solvent Effects on pKa values of Some Substituted Sulfonamides in Acetonitrile-Water Binary Mixtures by the UV-Spectroscopy Method. Journal of Chemical and Engineering Data 2009;54 3014-3021.

[7] Białk-Bielińska A., Stolte S., Matzke M., Fabiańska A., Maszkowska J., Kołodziejska M., Liberek B., Stepnowski P., Kumirska J. Hydrolysis of sulphonamides in aqueous solutions. Journal of Hazardous Materials 2012.; doi:10.1016/j.jhazmat.2012.04.044.

[8] Stoob K. Veterinary sulfonamide antibiotics in the environemnt: fate in grassland soils and transorpt to surface waters. PhD thesis. Swiss Federal Institute of Technology Zurich, Zurich; 2005.

[9] http://www.vcclab.org/lab/alogps/start.html (ALOGPS 2.1 program)

[10] http://www.syrres.com/what-we-do/databaseforms (SRC PhysProp Database)

[11] Babić S., Horvat A. J. M., Mutavdžić Pavlović D., Kaštelan-Macan M. Determination of pKa values of active pharmaceutical ingredients. Trends in Analytical Chemistry 2007;26 1043-1061.

[12] Carda-Broch S., Berthod A. Countercurrent chromatography for the measurement of the hydrophobicity of sulfonamide amphoteric compounds. Chromatographia 2004;59 79-87.

[13] Ruiz-Angel MJ., Carda-Broch S., García-Alvarez-Coque MC., Berthod A. Effect of ionization and the nature of the mobile phase in quantitive structure-retention relationship studies. Journal of Chromatogrphy A 2005;1063 25-34.

[14] Kümmerer, K. Antibiotics in the aquatic environment – A review – Part I, II. Chemosphere 2009;75 417-434.

[15] Petrović M., Barceló D., editors. Analysis, fate and removal of pharmaceutical in the water cycle. Comprehensive Analytical Chemistry. Amsterdam: Elsevier; 2007.

[16] Park S., Choi K. Hazard assessment of commonly used agricultural antibiotics on aquatic ecosystems. Ecotoxicology 2008;17 526-538.

[17] García-Galán MJ., Díaz-Cruz MS., Barceló D. Combining chemical analysis and ecotoxicity to determine environmental exposure and assess risk from sulfonamides. Trends in Analytical Chemistry 2009;28 804-819.

[18] Schauss K., Focks A., Heuer H., Kotzerke A., Schmitt H., Thiele-Bruhn S., Smalla K., Wilke B.M., Matthies M., Amelung W., Klasmeier J., Schloter M. Analysis, fate and effects of the antibiotic sulfadiazine in soil ecosystems. Trends in Analytical Chemistry 2009;28 612-618.

[19] Molander L., Ågerstrand M., Rudén C. WikiPharma – A freely available, easily accessible, interactive and comprehensive database for environmental effect data for pharmaceuticals. Regulatory Toxicology and Pharmacology 2009;55 367-371.

[20] Santos LHLM., Araújo AN., Fachini A., Pena A., Delerue-Matos C., Montenegro, MCBSM. Ecotoxicological aspects related to the presence of pharmaceuticals in the aquatic environment. Journal of Hazardous Materials 2010;175 45-95.

[21] Carlsson C., Johansson AK., Alvan G., Bergman K., Kühler T. Are pharmaceuticals potent environmental pollutants? Part II: Environmental risk assessments of selected pharmaceutical excipients. Science of the Total Environment 2006;364 67-87.

[22] Brain RA., Johnson DJ., Richards SM., Hanson ML., Sanderson H., Lam MW., Young C., Mabury SA., Sibley PK., Solomon KR. Microcosm evaluation of the effects of an eight pharmaceutical mixture to the aquatic macrophytes Lemna gibba and Myriophyllum sibiricum. Aquatic Toxicology 2004;70 23-40.

[23] Brain RA., Ramirez AJ., Fulton BA., Chambliss CK., Brooks BW. Herbicidal effects of sulfamethoxazole in Lemna gibba: Using p-aminobenzoic acid as a biomarker effect. Environmental Science and Technology 2008;42 8965-8970.

[24] Grote M., Chwanke-Anduschus C., Michel R., Stevens H., Heyser W., Langenkämper G., Betsche T., Freitag M. Incorporation of veterinary antibiotics into crops from manured soil. Lanbauforschung Völkenrode 2007;1 25-32.

[25] De Liguoro M., Di Leva V., Gallina G., Faccio E., Pinto G., Pollio A. Evaluation of the aquatic toxicity of two veterinary sulfonamides using five test organisms. Chemosphere 2010;81 788-793.

[26] De Liguoro M., Fioretto B., Poltronieri C., Gallina G. The toxicity of sulfamethazine to Daphnia magna and its additivity to other veterinary sulfonamides and trimethoprim. Chemosphere 2009;75 1519-1524.

[27] Baran W., Sochacka J., Wardas W. Toxicity and biodegradability of sulfonamides and products of their photocatalytic degradation in aqueous solutions. Chemosphere 2006;65 1295-1299.

[28] Eguchi K., Nagase H., Ozawa M., Endoh YS., Goto K., Hirata K., Miyamoto K., Yoshimura H. Evaluation of antimicrobial agents for veterinary use in the ecotoxicity test using microalgae. Chemosphere 2004;57 1733-1738.

[29] Isidori M., Lavorgna M., Nardelli A., Pascarella L., Parrella A. Toxic and genotoxic evaluation of six antibiotics on non-target organisms. Science of the Total Environment 2005;346 87-98.

[30] Kim Y., Choi K., Jung J., Park S., Kim PG., Park J. Aquatic toxicity of acetaminophen, carbamazepine, cimetidine, diltiazem and six major sulfonamides, and their potential ecological risks in Korea. Environment International 2007;33 370-375.

[31] Pro J., Ortiz JA., Boleas S., Fernández C., Carbonell G., Tarazona JV. Effect assessment of antimicrobial pharmaceuticals on the aquatic plant *Lemna minor*. Bulletin of Environmental Contamination and Toxicology 2003;70 290-295.

[32] Tappe W., Zarfl C., Kummer S., Burauel P., Vereecken H., Groeneweg J. Growth-inhibitory effects of sulfonamides at different pH: Dissimilar susceptibility patters of a soil bacterium and a test bacterium used for antibiotic assays. Chemosphere 2008;72 836-843.

[33] Białk-Bielińska A., Stolte S., Arning J., Uebers U., Böschen A., Stepnowski P., Matzke M. Ecotoxicity evaluation of selected sulfonamides. Chemosphere 2011;85 928-933.

[34] U.S. Deparment of Health and Human Services, Food and Drug Administration. Guidance for Indrustry: Environmental Assessment of Human Drug and Biologics Application; July 1998; www.fda.gov/cder/guidance/1730fnl.pdf.

[35] U.S. Deparment of Health and Human Services, Food and Drug Administration. Guidance for Indrustry: Environmental Assessment for Veterniary Medical Products – Phase I; March 2001; www.fda.gov/cvm/Guidance/guide89.pdf.

[36] CHMP (Committee for Medical Products for Human Use), Guideline on the Environmental Risk Assessment of Medical Products for Human Use, EMEA/CHMP/SWP/4447/00, London 2006.

[37] CVMP (Committee for Medical Products for Veterinary Use), Revised Guideline on the Environmental Risk Assessment for Veterinary Medicinal Products in Support of the VICH GL6 and GL 38, EMEA/CVMP/ERA/418282/2005-Rev.1, London 2008.

[38] VICH (International Cooperation on Harmonization of Technical Requirements for Registration of Veterinary Medical Products), Guideline GL 6 on Environmental Impact Assessment (EIAs) for Veterinary Medicinal Products - Phase I, CVMP/VICH/592/98-FINAL, London 2000.

[39] VICH (International Cooperation on Harmonization of Technical Requirements for Registration of Veterinary Medical Products), Guideline GL 38 on Environmental Impact Assessment for Veterinary Medicinal Products - Phase II, CVMP/VICH/790/03-FINAL, London 2005.

[40] Schmitt H., Boucard T., Garric J., Jensen J., Parrott J., Péry A., Römbke J., Straub JO., Hutchinson TH., Sánchez-Argüello P., Wennmalm Å., Duis K. Recommendations on the Environmental Risk Assesment of Pharmaceuticals – effect characterization. Integrated Environmental Assessment and Management 2010;6 588-602.

[41] Tarazona JV., Escher BI., Giltrow E., Sumpter J., Knacker T. Targeting the environmental risk assessment of pharmaceuticals: fact and fantasies. Integrated Environmental Assessment and Management 2010;6 603-613.

[42] Ankley GT., Brooks BW., Huggett DB., Sumpter JP. Repeating history: Pharmaceuticals in the environment. Environmental Science and Technology 2007;15 8211-8217.

[43] Jonker MJ., Svendsen C., Bedaux JJM., Bongers M., Kammenga JE. Siginificance testing of synergisitic/antagonistic, dose level-dependent, or dose ratio-dependent effects in mixture dose-response analysis. Environmental Toxicology and Chemistry 2005;24 2701-2713.

[44] Loewe S., Muischnek H. über Kombinationswirkungen. 1. Mitteilung: hilfsmittel der fragestellung. Naunyn-Schmiedebergs Archiv für Experimentelle Pathologie und Pharmakologie 1926;114 313-326.

[45] Bliss CI. The toxicity of poisons applied jointly. Annual Review of Applied Biology 1939;26 585-615.

[46] Cleuvers M. Aquatic ecotoxicity of pharmaceuticals including the assessment of combination effects. Toxicology Letters 2003;142 185-194.

[47] Cleuvers M. Mixture toxicity of the anti-inflamatory drugs diclofenac, ibuprofen, naproxen and acetylsalicylic acid, Ecotoxicology and Environmental Safety 2004;59 309-315.

[48] Berenbaum MC. The expected effect of a combination of agents: the general solution. Journal of Theoretical Biology1985;114 413-431.

[49] Altenburger R., Backhaus T., Boedeker W., Faust M., Scholze M., Grimme LH. Predictability of the toxicity of multiple chemical mixtures to *Vibrio fischeri*: mixtures composed of similarly acting chemicals. Environmental Toxicology and Chemistry 2000;19 2341-2347.

[50] Backhaus T., Altenburger R., Boedeker W., Faust M., Scholze M., Grimme LH. Predictability of the toxicity of a multiple mixture of dissimilarly acting chemicals to *Vibrio fischeri*. Environmental Toxicology and Chemistry 2000;19 2348-2356.

[51] Faust M., Altenburger R., Bachaus T., Blanck H., Boedeker W., Gramatica P., Hamer V., Scholze M., Vighi M., Grimme LH. Predicitng the joint algal toxicity of multicomponenet s-triazine mixtures at low-effect concentrations of individual toxicants. Aquatic Toxicology 2001;56 13-32.

[52] Cassee FR., Groten JP., van Bladeren PJ., Feron VJ. Toxicological Evaluation and Risk Assessment of Chemical Mixtures, Critical Reviews in Toxicology 1998;28(1) 73–101.

[53] Van Loon WMGM., Verwoerd ME., Eijnker FG., van Leeuwen CJ., van Duyn P., van deGuchte, Hermens JLM. Estimating total body residues and baseline toxicity of complex organic mixtures in effluents and surface waters. Environmental Toxicology and Chemistry 1997;16 1358-1365.

[54] Verhaar HJM., van Leeuwen CJ., Hermens JLM. Classifying environmental pollutants. Chemosphere 1992;25 471-491.

[55] Crane M., Watts C., Boucard T. Chronic aquatic environmental risks from exposure to human pharmaceuticals. Science of the Total Environment 2006;367 23–41.

[56] Waller WT., Allen HJ. Acute and Chronic Toxicity. Ecotoxicology 2008; 32-43.

[57] Blasco J., DelValls A. Impact of emergent contaminant in the environment: Environmental Risk Assessment. Handbook of Environmental Chemistry 2008;5 169-188.

[58] Heuer H., Smalla K. Manure and sulfadiazine synergistically increased bacterial antibiotic resistance in soil over at least two months. Environmental Microbiology 2007;9(3) 657-666.

[59] Monteiro SC., Boxall ABA. Factors affecting the degradation of pharmaceuticals in agricultural soils. Environmental Toxicology and Chemistry 2009;28(12) 2546-2554.

[60] Wollenberger L., Halling-Sørensen B., Kusk KO. Acute and chronic toxicity of veterinary antibiotics to *Daphnia magna*. Chemosphere 2000;40 723-730.

[61] Zou X., Lin Z., Deng Z., Yin D., Zhang Y. The joint effects of sulfonamides and their potentiator on *Photobacterium phosphoreum:* Differences between the acute and chronic mixture toxicity mechanisms. Chemosphere 2012;86 30-35.

[62] Kümmerer K., Alexy R., Hüttig J., Schöll A. Standardized tests fail to assess the effects of antibiotics on environmental bacteria, Water Research 2004;38 2111-2116.

[63] Backhaus T., Grimme LH. The toxicity of antibiotic agents to the bioluminescent bacterium *Vibrio fischeri*. Chemosphere 1999;38 3291-3301.

[64] Bolelli L., Bobrovová Z., Ferri E., Fini F., Menotta S., Scandurra S., Fedrizzi G., Girotti S. Bioluminescent bacteria assay of veterinary drugs in excreta of food-producing animals. Journal of Pharmaceutical and Biomedical Analysis 2006;42 88-93.

Chapter 4

Perfluorinated Organic Compounds and Polybrominated Diphenyl Ethers Compounds – Levels and Toxicity in Aquatic Environments: A Review

Monia Renzi

Additional information is available at the end of the chapter

1. Introduction

Organochlorine pesticides are well known by the scientific literature to be persistent in the environment, toxic for wildlife and potential dangerous for humans since the publication of the famous volume "Silent Spring" by Rachel Carson in 1962 [1]. Due to their massive industrial production and commercialization for various human purposes, these chemicals have reached concentrations in worldwide environments that are able to significantly affect terrestrial and marine wild species [2], remote world habitats [3] including the remote deep-sea [4], and protected species listed in the IUCN Red List of Threatened Species [5]. Actual concentrations are able to severely affect trophic webs [6] and top-predators [7; 8]. As animal, humans are not excluded by the effects of pollution and concerning these compounds, feeding represents the principal and worldwide diffuse exposure mechanism for human populations rather than inhalation and dermal contact [9; 10; 11; 12]. In spite of that, only in USA, chemical industry produces about 70,000 new products and organic chemicals accounted for the greatest share of production (364.2 million tons) in 1997 [13].

The continuous research of new substances able to cover the great request from engineering, chemistry, pharmaceuticals, medical, commercial and social activities led to the direct and indirect release in environment and the consequent exposure of living organisms to new compounds. Once released in the environment, new chemicals interact with the abiotic and biotic matrices producing mixture composed by pure chemicals which auto-interact, their metabolites and/or reaction and degradation by-products. These mixture are characterized by a progressive increase of complexity and by a clear geographical footprint with percentages of chemical composition that are dependent both by physico-chemical properties of compounds making the mixture and latitude/altitude of the geographical area considered [14]. Unluckily the effects induced by pure compounds on non-target species are frequently unknown at the time of their commercialization as well as possible by-products which are produced by the interaction with the environment. Usually, undesirable consequences of new synthesized chemicals are discovered many years later their distribution in commerce, often dramatically. This is the well-known case of the pesticide dichloro-diphenyl-trichloro-ethane (DDT) largely used to control malaria diffusion and publicized before 1970' as "the best friend of housewives in controlling pests".

Persistent organic pollutants (POPs) are characterized by molecular stability, high persistence due to the resistance to natural degradation processes derived by physical (i.e. temperature or photo-degradation), chemical (i.e. redox and acid-basic reactions, chemical interactions), and biological (i.e. metabolic or microbial deteriorations) aggressions. As reported by the European Community [15], to be classified as "persistent", chemicals must evidence a half-life in water superior than two months and in sediments/soils superior than six months.

POPs concentrate in environment for a very long time and, due to their vapor pressure <1000 Pa and a half-life >2 days in atmosphere, evidence long range transport reaching, also, remote areas [16]. These chemicals usually are low water soluble but evidence a great affinity towards lipids and tend to accumulate in sedimentary organic matter and biological tissues affecting the trophic web along which tend to biomagnificate [17]. Chemicals characterized by $\log K_{ow} > 5$ and by a bio-concentration factor (BEF) >5,000 are considered "bioaccumulable" [15]. POPs are not biologically inert, on the contrary, they actively interact with physiological biochemistry of species inducing toxicity on wildlife species and humans.

Among POPs, perfluorinated organic compounds (PFCs) and polybrominated diphenyl ethers (PBDEs) are known as "emergent pollutants". PFCs and PBDEs are recently commercialized chemicals of particular ecotoxicological concern which are relatively little described by the literature [18]. PFCs and PBDEs increased levels during the latest decades both in environments and wildlife. Several studies have assessed them in a wide range of organisms [19], including humans [20; 21], from low latitude regions to remote areas, suggesting atmospheric transport of volatile precursor compounds and/or transport in ocean currents [22; 23; 24].

This chapter will focuses:

- general physico-chemical properties,
- sources, distribution dynamics, and environmental levels (in air, soil, water, sediment) with a particular attention on aquatic ecosystem;
- levels in wildlife tissues focusing evidences of bioaccumulation throughout the trophic web. Studies reporting levels both in red-list included species and foods at the basis of the human diet will be considered and included;
- phenomena of contamination in humans;
- evidences on toxicity based on results of ecotoxicological tests;
- international normative and guidelines developed to control considered chemicals

The following paragraphs aims to summarized actual knowledge on PFCs and PBDEs principal characteristics including environmental levels and toxicity on biota.

2. Physico-chemical properties of considered molecules

PFCs and PBDEs include molecules characterized by a similar chemical formula but also by very different physico-chemical properties. As consequence of the structural dissimilarities, differences concerning environmental distribution dynamics, and levels in abiotic and biological matrices are observed among PFCs and PBDEs congeners. Furthermore, the ecotoxicological risk associated to the diffusion of these persistent organic pollutants could be notably dissimilar. In fact, physico-chemical properties of molecules influence possible adverse effects on non-target biological communities. In addition, observed toxicity is notably affected by the interaction among considered chemicals and environmental matrices caused by the photo-chemical deterioration and the production of metabolites during microbial biodegradation phenomena.

2.1. Perfluorinated organic compounds (PFCs)

Concerning chemicals of ecotoxicological interest, perfluorinated organic compounds (PFCs) are of particular emerging interest due their documented presence both in wildlife's tissues and human blood PFCs [25].

PFCs are anionic, and fluorine-containing surfactants (both soluble in water and oil) and are applied for a large industrial and commercial employment to produce surfactants, lubricants, paints, polishes, food packaging, and fire-fighting, foams propellants, agrochemicals, adhesives, refrigerants, fire retardants, and medicines [26; 27].

Their structure consisting of a fluorine atom with which all hydrogen atoms from the linear-alkyl chain, which is a hydrophobic group, are replaced. Physico-chemical properties of PFCs favour the occurrence of long-range transport dynamics, as they are more volatile than chlorine or bromine analogues.

Among PFCs, perfluorooctanoic acid (PFOA) and perfluoroctanesulfonic acid (PFOS) represents the principal compounds of environmental concern.

Salts of perfluorooctanoic acid (PFOA, $C_8HF_{15}O_2$) have been used as surfactants and processing aids in the production of fluoropolymers, and these salts are considered critical to the production of certain fluoropolymers and fluoroelastomers [28]. The functional chemical structure is $C_7F_{15}COOH$ and for this reason tends to behavior like an acid dissociating as follows: $C_7F_{15}COO^- + H^+$.

Perfluorooctane sulfonate (PFOS; $C_8HF_{17}O_3S$ even in this case it dissociates as follows: $C_7F_{17}SO_3^- + H^+$) evidences an excellent chemical and thermal stability and is a chemical precursor for the synthesis of other molecules [26] such as fluorinated surfactants and pesticides (Abe and Nagase., 1982 in [29]). Perfluoroalkanesulfonate salts and perfluorocarboxylates are reported to be present in fire-fighting foam formulations, including aqueous film forming foams which are fire-fighting materials largely used by military bases and airports to face hydrocarbon fuel fires or to prevent the potential risk of fire [30; 31]. Moody and colleagues [32] reported for the 2001, an estimated PFOS annual production quantity in United States of America of 2,943,769 kg.

Principal PFOA and PFOS chemical properties are summarized in Table 1.

	PFOA	PFOS
Extended name	Perfluorooctanoic acid	Perfluoroctanesulfonic acid
Other names	Perfluorooctanoate Perfluorocaprylic acid FC-143 F-*n*-octanoic acid	1-Perfluorooctanesulfonic acid Heptadecafluoro-1-octanesulfonic acid Perfluoro-n-octanesulfonic acid
	SUBSTANCE IDENTIFICATION	
CAS numb	335-67-1	1763-23-1
Pubchem	9554	74483
EC number	206-397-9	217-179-8
	MOLECULAR PROPERTIES	
Molecular formula	$C_8HF_{15}O_2$	$C_8HF_{17}O_3S$
Molecular mass	414.07 $gmol^{-1}$	500.13 $gmol^{-1}$
Boiling point	189–192 °C	133 °C (6 torr)
Appearance (25 C, 100 kPa)	colorless liquid	white powder
Vapor pressure	4.2 Pa (25 °C)	3.31 x 10-4 Pa (20 °C)
Melting point	40–50 °C	>400 °C
Solubility in water	3,400 mgL^{-1}	519 mgL^{-1} (20 ± 0.5 °C) 680 mgL^{-1} (24 - 25 °C)
Solubility in other solvents	polar organic solvents	56 mgL^{-1} (octanol)
Acidity (pKa)	2-3[23]	calculated value of -3.27[33]
	RELATED RISKS	
S-phrases	S36, S37, S39	S61
R-phrases	R22, R34, R52/53	R61, R20/21, R40, R48/25, R64, R51/53

Table 1. Substance identification (extended names and international classification numbers), principal molecular properties, and related risks of PFOA (perfluorooctanoic acid) and PFOS (perfluoroctanesulfonic acid) are summarized in table. Specific references: record of PFOA were extracted from the GESTIS Substance Database from the IFA (last access on 5th November, 2008).

2.2. Polybrominated Diphenyl Ethers (PBDEs)

Polybrominated diphenyl ethers (PBDEs) are a class of organohalogen compounds used worldwide over the past three decades as chemical additives to reduce the flammability of common use products [34]. These chemicals were first introduced to the market in the 1960s and their global demand has increased rapidly since the end of the 1970s, due to the growing popularity of personal computers and other electronic equipment, to which they were added to improve fire safety [35].

Since '70 PBDEs were used as flame retardants in a wide range of common use such as cloths, foam cushions, polyurethane sponges, carpet pads, chairs, couches, electronic instruments including computer castings, and insulating materials [36].

In 2000, the industrial production of these chemicals has been esteemed to be around the 64,000 cubic tons per year. The 50% of this annual production was commercialized in America, while in Europe only 12% [37].

Because of toxicity and persistence of PBDEs, these chemicals are included in the persistent organic pollutants (POPs) Reviewing Committee (www.pops.int) and their industrial production is to be eliminated under the Stockholm Convention.

Polybrominated diphenyl ethers are, apart from the oxygen atom between the phenyl rings, structurally similar to PCBs, consisting of two halogenated aromatic rings linked by an ether group. PBDEs chemical synthesis is performed by the diphenyl-ethers bromination in presence of dibromomethane as solvent. Diphenyl-ethers have 10 hydrogen atoms and each of them can be replaced by an atom of bromine. This reaction could produce 209 possible congeners, numbered from 1 to 209 in relation to the number of bromine atoms substituting hydrogen ones and their relative position within the molecule [38].

The general chemical formula for PBDE family is $C_{12}H_{(10-x)}Br_xO$ (where x= 1,..., 10).

In the United States, PBDEs are marketed with trade names: DE-60F, DE-61, DE-62, and DE-71 applied to penta-BDE mixtures; DE-79 applied to octa-BDE mixtures; DE 83R and Saytex 102E applied to deca-BDE mixtures.

The available commercial PBDE products are not single compounds or even single congeners but rather a mixture of congeners. Nevertheless, commercial mixtures are constituted by a little part of the 209 possible congeners due to the instability of a large part of them [39] which tend to quickly debrominate.

Three technical mixtures are available and commercialized and differ related to the bromination levels:

- Mixture penta-BDE (24-38% tetra-BDE, 50-60% penta-BDE, 4-8% esa-BDE). In these mixtures, most abundant congeners are constituted by tetra-BDE 2,2′,4,4′ (IUPAC n. 47), penta-BDE 2,2′,4,4′,5 (IUPAC n. 99) and penta-BDE 2,2′,4,4′,6 (IUPAC n. 100), esa-BDE 2,2′,4,4′,5,5′ (IUPAC n. 153) and esa-BDE 2,2′,4,4′,5,6′ (IUPAC n. 154). These mixtures are viscose liquids principally used in industrial fabrication of clothes, foams, resins, polyurethane foam products such as furniture and upholstery in domestic furnishing, and in the automotive and aviation industries. The European Union banned the use of this mixture in August 2004.

- Mixture octa-BDE (10-12% esa-BDE, 44% epta-BDE, 31-35% octa-BDE, 10-11% nona-BDE, <1% deca-BDE). In these mixtures, most abundant congeners are epta-BDE 2,2′,4,4′,5′,6 (IUPAC n. 183), and esa-BDE 2,2′,4,4′,5,5′ (IUPAC n. 153). These mixtures are white dusts and are commonly used in little objects for house and office purposes made by plastic products, such as housings for computers, automobile trims, telephone handsets and kitchen appliance casings.
- Mixture deca-BDE (<3% nona-BDE, >97% deca-BDE (IUPAC n. 209). These mixtures are white dusts. In 2003 they represent above the 80% of the annual production of PBDE and they are, currently, the only PBDE product in production. Deca-BDE are commonly used in the following applications: thermoplastic, elastomeric, and thermo set polymer systems, including high impact polystyrene (HIPS), polybutylene terephthalate (PBT), nylon, polypropylene, low-density polyethene (LDPE), ethylene-propylene-diene rubber and ethylene-propylene terpolymer (EPDM), unsaturated polyester, epoxy. Are used for wire and cable insulation, coatings and adhesive systems, including back-coatings for fabrics, and electronic instruments [36; 38].

PBDEs are semi volatile compounds characterized by a low vapor pressure and a scarce water solubility. These properties tends to decrease with the level of substitutions by bromine atoms in the molecular structure whereas hydrophobic properties increase. Octanol/water distribution coefficients (K_{ow}) are variable with substitutions and are included within: 5.9-6.2 for tetra-BDE, 6.5-7.0 for penta-BDE, 8.4-8.9 for octa-BDE, and 10.0 for deca-BDE. PBDEs half-life in air are extimated to be about two days, while in water longer times are modeled (two months) whereas in soils and sediments average half-lives are six months [40].

PBDEs	Isomer	Molecular formula	Molecular mass	% bromine	Vapor pressure	Octanol/water distribution coefficient	Solubility in water
Measure ment unit	-	-	*g/mol*	*m/m*	*25°C, Pa*	*log Pow*	*21 °C, µg/L*
mono-PBDE	3	$C_{12}H_9BrO$	249.0	32.09		3.6	4000
di-PBDEs	12	$C_{12}H_8Br_2O$	327.9	48.74	2.0	5.1	500
tri-PBDE	24	$C_{12}H_7Br_3O$	406.8	58.93	$2.0\ 10^{-2}$	5.9	90
tetra-PBDE	42	$C_{12}H_6Br_4O$	485.7	65.81	$4.0\ 10^{-4}$	6.3	20
penta-PBDE	46	$C_{12}H_5Br_5O$	564.6	70.77	$3.0\ 10^{-5}$	6.8	5
hexa-PBDE	42	$C_{12}H_4Br_6O$	643.5	74.51	$9.0\ 10^{-6}$	7.3	2
hepta-PBDE	24	$C_{12}H_3Br_7O$	722.4	77.43	$5.0\ 10^{-6}$	7.9	0.7
octa-PBDE	12	$C_{12}H_2Br_8O$	801.3	79.78	$4.0\ 10^{-6}$	8.5	0.3
nona-PBDE	3	$C_{12}HBr_9O$	880.1	81.71	$3.0\ 10^{-6}$	9.0	0.16
deca-PBDE	1	$C_{12}Br_{10}O$	959.0	83.32	$2.6\ 10^{-6}$	9.5	0.10

Table 2. Substance identification and principal molecular properties of PBDEs (polybrominated diphenyl ethers) are summarized. The number of isomers, the molecular formula, molecular mass, % of bromine, vapor pressure, octanol/water distribution coefficients, and solubility in water are reported. Data collected by the European Food Safety Authority [41].

3. Sources, distribution dynamics, and environmental levels

Concerning PFCs, principal environmental sources are represented by the direct diffusion of surfactants, lubricants, paints, polishes, foams propellants, agrochemicals, adhesives, refrigerants, fire retardants, and medicines containing these chemicals. Indirect releases could occurs from food packaging and painted manufacturing when discharged and exposed to rain and bad weather conditions. Nevertheless, the large use of fire-fighting materials containing PFCs both when a critical fire occurs and to prevent accidents in high risk procedure (i.e. military or firemen exercitations, routine activities, airports activities), represents the principal direct diffusion of these chemicals on the ground able to affect wide geographical surfaces, superficial and groundwater [31].

As regard as PBDEs, environmental releases could occurs during manufacturing lifetimes. Releasing mechanism are not completely cleared, however, it is believed that PBDEs are released to the air when objects are manufactured and during object's life span. Their disposal and waste could produce releases too [42]. In the last years recycling of end products containing PBDE is becoming the principal source of release of these chemicals in the environment [43]. Burning of plastics, waste electronic goods, and oil shale may provide an additional PBDEs loads both in atmosphere and soil. Also, productive processes represents an important source, high levels are measured in environmental matrices closed to the flame-retardants factories [40].

Monitoring PFCs and PBDEs in environmental matrices evidenced first of all the needing to develop accurate sampling strategies to collect representative samples from heterogeneous and quickly variable matrices such as air and water are. On the contrary, soils and sediments even if much more stable present structural heterogeneity (i.e. organic matter content and composition, grain-size structure, redox conditions) which could interfere with quantifications and data interpretation. Concerning biota the matter (if it is possible!) is quite more complex. Measured levels could be affected by a lot of different factors as well as age, sex, phase of animal life-stage, lipid content, water content in tissues, part of the animal excised for the analyses and much more other factors. Another point is represented by the sampling treatments and the detecting method adopted to perform laboratory analyses. Different methods are associated to different detection limits, precision and accuracy. Low polluted matrices such as air and water required methods able to detect levels of chemicals at concentrations measured in pg/L, whereas biological tissues allowed the adoption of quite less sensible methods as well as concentrations are usually measured in mg/kg or ng/g.

Hereby levels reported by the literature in different environmental matrices are reported organizing them per matrix. When possible information about the sampling strategies adopted are reported (i.e. depth of sampling for water and soils or sediments, geographical areas, type of tissues), nevertheless a complete data selection related to the sampling strategies, sampling treatments, and detecting methods has not been possible due to the wide heterogeneity in data acquisition procedures.

Extremely summarizing, perfluorooctanoic acid (PFOA) is dominant in environmental matrices whereas perfluorooctane sulfonates (PFOS) represents the predominant compound found in biota [44].

3.1. Air

Low data are available on air levels, probably due to the great difficult associated to the sampling of this matrix and samples treatment strategies in laboratory. Laboratory (i.e. air, laboratory rooms, instruments, vials, etc.) and cross-over contaminations are extremely simple to occur treating air samples. Furthermore, adopted methods have to be extremely sensitive.

In the period from 1994 to 1995, measured levels of total PBDEs (congeners not specified) reached maximum values of 28 pg/m^3 in samples collected in Alert, Canadian Arctic [36]. The rural area of southern Ontario showed in the Early spring of 2000, notably higher levels of total PBDEs (as sum of 21 congeners detailed in the paper) ranging within 10-1,300 pg/m^3 [45].

Samples collected in Great Lakes from 1997-1999 evidenced total PBDEs ranging within 5.5-52 pg/m^3 [46], comparable levels (3.4-46 pg/m^3) were measured in Ontario (2000) by Harner and colleagues [47].

3.2. Terrestrial environments and soils

Soil pollution could derived by direct local sources but also by dry-air depositions or runoffs. Humus represents the soil fraction able to accumulate chemicals due to the presence of both hydrophobic and hydrophilic molecules. From here chemicals could be re-volatilized in air, transferred throughout the soil trophic web or be leached throughout rains affecting groundwater ecosystems with possible important consequences for humans. The net dominance of one of these phenomena is a factor dependent to the geographical position of the area (affecting air/soil temperature, sun irradiance, quantities of rains, etc.) and to the soil physico-chemical characteristics.

In soils sampled closed to a polyurethane foam manufacturing facility in the United States, concentrations of total PBDEs (tetra- and penta-BDE) of 76 μg/kg dry weight are reported. Average values measured in soil downwind from the facility were significantly lower 13.6 μg/kg dry weight [48; 49].

In a study performed throughout the Estonian State, PBDEs levels are defined in soils for the first time. Total values observed ranging within <0.01-3.2 ng/g (d.w.) as reported by Kumar and colleagues [50]. Even if measured values are not excessive, authors predict a possible increase in the near future due to the particular waste policy of the Estonia.

In Australia, superficial samples collected in 39 remote, agricultural, urban and industrial locations from all states and territories evidenced highest (but not indicated) values from "*urban and industrial areas, particularly downstream from sewage treatment plants*" (Australian Government, on-line available at: http://www.environment.gov.au/settlements/publications/

chemicals/bfr/pubs/factsheet.pdf.). In the same report, levels are indicated to be comparable to values measured in other European and Asian Countries.

3.3. Aquatic environments

Aquatic ecosystems represent the final reservoir for PFCs and PBDEs due to their great affinity towards sedimentary and living organic matter. In these systems, measured levels of POPs could increase along the trophic web affecting humans feeding aquatic species.

A recent study performed by Nakata and colleagues [51] evidenced significant differences among tidal and coastal levels of PFOA and PFOS in all considered environmental matrices supporting the existence of different dynamics affecting PFCs distribution and ingress in trophic webs that are zone dependant in marine ecosystems.

Even if some researchers, as reported below, have been performed to evaluate PFCs and PBDEs levels in environmental matrices and biota from river and marine ecosystems, no data are available on environmental levels, bioaccumulation and biomagnification dynamics occurring in coastal lagoons and transitional areas which are completely non explored. This lacking in scientific data could affect risk evaluations linked to human exposure to these chemicals in transitional areas. In fact lagoons and estuaries are the most populated, polluted and productive areas in the world. Feeding exploitation of these not explored ecosystems could represent a notable and not considered risk for humans.

3.3.1. Water

A report produced by IFA [52] documented PFOAs levels in drinking and surface fresh water (n=440) ranging within 0.05-456 ng/L. In Europe (n=119) levels recorded are included within 0.33-57.0 ng/L. On the contrary, in the same dataset, measured PFOS levels in drinking and surface fresh water range within 7.1-135 ng/L, while, in Europe values are included within 21.8-56.0 ng/L reporting minimum values notably higher than PFOA levels.

Data acquired on PFOS levels from Six U.S. Urban Centres [53] evidenced ranges within <0.01-0.063 (ppb) in drinking water and values included within 0.041-5.29 (ppb) in Municipal wastewater treatment plant effluents (MWTP). Surface water are included within <0.01-0.138 (ppb) while "quiet" water values are similar to those observed in MWTP (<0.01-2.93 ppb). These data evidenced that treatment plant effluents could contribute significantly to superficial watercourse pollution.

In 2004, Boulanger and colleagues [54] explored for the first time perfluorooctane surfactants concentrations in sixteen Great Lakes water also determining PFOA and PFOS precursors samples from Great Lakes. Levels measured in water ranged within 27-50 ng/L (PFOA) and 21-70 ng/L (PFOS). The presence of PFOS precursors was recorded in all samples above the LOQ.

Hansenk and colleagues [55] performed a monitoring of the superficial Tennessee River water to evaluate possible contribution to water levels due to the activity of a fluorochemical manufacturing site (settled in Decatur, AL). PFOA levels reported are always below the detection limit (25 ng/L) with the exception of samplings collected closed to the fluorochemical plant where PFOA values ranged within 140-598 ng/L. PFOS are recorded at low but often measurable levels (<25-52 ng/L) in river sampling stations evidencing a significant increase closed to the fluorochemical manufacturing facility (74.8-144.0 ng/L). This research suggests a strong contribution of plant's outflows to river PFOA and PFOS levels.

In 2005, Yamashita and colleagues [56] developed a reliable and highly sensitive analytical method to monitor PFCs in oceanic water. Between 2002-2004, levels measured in Pacific Ocean (n=19), South China Sea and Sulu Seas (n=5), north and mid Atlantic Ocean (n=12), and the Japan Sea (n=20) were respectively of: 15-142 pg/L, 76-510 pg/L, 100-439 pg/L, 137-1,070 pg/L for PFOA and 1.1-78 pg/L, <17-113 pg/L, 8.6-73 pg/L, and 40-75 pg/L for PFOS. Concerning PFOA, samples collected along coastal seawater from several Asian countries (Japan, China, Korea) evidenced levels included within: 1,800-19,200 pg/L (Tokyo Bay), 673-5,450 pg/L (Hong-Kong), 243-15,300 pg/L (China), and 239-11,350 pg/L (Korea). On the contrary, concerning PFOS values in the same sampling sites were: 338-57,700 pg/L (Tokyo Bay), 70-2,600 pg/L (Hong-Kong), 23-9,680 pg/L (China), and 39-2,530 pg/L (Korea).

A research performed in 2007 by Senthilkumar and colleagues [44] evidenced in Japan water PFOA concentrations of 7.9–110 ng/L and PFOS values ranging within <5.2–10 ng/L.

In 1999, PBDEs levels (mono- to hepta-BDE congeners) concentrations of approximately 6 pg/L were measured in Lake Ontario surface waters [57]. In this study, more than 60% of the total was composed of BDE47 (tetra-BDE) and BDE99 (penta-BDE), with BDE100 (penta-BDE) and BDEs 153 and 154 (hepta-BDE congeners) each contributing approximately 5% to 8% of the total.

Stapleton and Baker [58] analyzed water samples from Lake Michigan in 1997, 1998 and 1999 founding total PBDEs concentrations (BDEs 47, 99, 100, 153, 154 and 183) ranging within 31-158 pg/L.

3.3.2. Sediment

In Japan aquatic environments, Senthilkumar and colleagues [44] observed PFOA measurable levels only in sediments sampled from the Kyoto river ranging within 1.3–3.9 ng/g dry weight (dw) and not measurable PFOS levels.

Becker and colleagues [59] evidenced that once released in water, PFCs accumulate into sediments with a PFOA/PFOS ratio of about 10. In particular, PFOA were 10-fold less than PFOS but enrichment observed on sediment was not correlated to the total organic carbon contents.

In 1998, Lake Michigan evidenced average values of total PBDE of 4.2 μg/kg dw [58].

Concerning PBDEs, levels measured in sediments from taken from fourteen Lake Ontario tributary sites [60] evidenced total PBDEs (tri-, tetra, penta-, hexa-, hepta- and deca-BDEs) levels ranging within 12-430 μg/kg dry weight, with tetra- to hexa-BDEs sum ranging within 5-49 μg/kg dry weight. Concentrations of BDE 209 ranged from 6.9 to 400 μg/kg dw and BDE 47, 99 and 209 were the predominant congeners measured in sediments.

From several sites sampled along the Columbia River system, in south eastern British Columbia, Rayne and colleagues [61] measured PBDE concentrations (as sum from di- to penta-BDE congeners) included within 2.7-91 μg/kg.

Sediments from two Arctic lakes in Nunavut Territory evidenced measurable concentrations from 0.075 to 0.042 μg BDE 209/kg dw. One of the two Arctic lakes sampled was located near an airport and PBDEs inputs from this source could not be excluded [62]. Authors hypothesized a particles-mediated transport to the Canadian Arctic due to its low vapour pressure and high octanol-water partition coefficient.

Sludge sampled from Municipal wastewater treatment plants evidenced total PBDEs (21 mono- to deca-BDE congeners) ranging from 1,414 to 5,545 μg/kg dw [63]. A regional sewage treatment plant discharging to the Dan River in Virginia evidenced in 2000 total PBDEs (sum of BDEs 47, 99, 100 and 209) of 3,005 μg/kg dw [48].

3.4. Biota

POPs could accumulate in species evidencing interspecies differences as well as sex and size-related ones [64]. Recent studies evidenced that POPs concentrations in demersal fishes varies significantly relating to the sex, maturity, and reproduction [65].

Data collected in fishes and fishery products [52] evidence PFOA levels ranging within 0.05-5.00 ng/g wet weight (w.w.) (muscle of whole body) in Europe [66], 0.13-18.70 ng/g w.w. in Asia [67; 68], and 0.70-2.40 ng/g w.w. in North America [69].

Crustaceans levels are quite similar in their edible parts and respectively of 0.80-0.90 ng/g w.w. [66], 0.13-9.50 ng/g w.w. [51], and 0.10-0.50 ng/g w.w. [70] respectively in Europe, Asia, and North America. Observed levels in edible part of molluscs ranged within 0.95-1.20 ng/g w.w. in species from Europe [66], 0.10-22.90 ng/g w.w. from Asia [67; 68]. Molluscs in North America showed levels closed to the detection limits (0.10 ng/g w.w.) as reported by Tomy and colleagues [70].

Concerning PFOS in fish muscles or whole body ranged within 0.60-230 ng/g w.w. in Europe, 0.380-37.30 ng/g in Asia [68], and 15.1-410 ng/g in North America [71]. Crustaceans evidences levels included within 8.30-319 in Europe [66], 0.15-13.9 in Asia [67], and 0.03-0.90 in North America [70], while molluscs showed PFOS levels of 0.80-79.80 in Europe [72], 0.114-47.200 in Asia [68], and 0.080-0.600 in North America.

Kannan and colleagues [73] performed a screening of PFOA and PFOS in wildlife species from different trophic levels and ecosystems. Concerning PFOS in blood samples collected in aquatic mammals and fishes, a tendency to the decrease of measured levels is reported for bottlenose dolphins>bluefish tuna>swordfish. They reported that PFOS concentrations (61 ng/g, w.w.) measured in cormorant livers collected from Sardinia Island (Italy) are lower than PFOA (95 ng/g, w.w.) but significantly correlated.

In the same research, PFOS levels measured in liver samples collected from ringed and gray seals (Bothnian Bay, Baltic Sea) range within 130-1,100 ng/g, w.w.. In this case, no relationships are observed between PFOS levels and ringed or gray seals age but levels measured in livers are 2.7-5.5 fold higher than values in blood with a positive strong correlation between blood and liver levels. Concerning white-tailed sea eagles (Germany and Poland) indicate increasing of concentrations from 1979 to 1990s. Livers of Atlantic salmons do not evidenced measurable levels neither PFOS nor PFOA.

In 2007, Senthilkumar and colleagues [44] define levels of PFOA and PFOS in biotic compartment of aquatic ecosystems in Japan. Concerning fish tissues, only jack mackerel showed PFOA and PFOS respectively at averages of 10 and 1.6 ng/g w.w.. Wildlife livers contained PFOS levels ranging within 0.15–238 ng/g w.w. and PFOA values included within <0.07–7.3 ng/g w.w.. Cormorants showed maximum accumulation followed by eagle, raccoon dog and large-billed crow.

Kannan and colleagues [73] measured PFOA and PFOS levels in livers of birds collected from Japan and Korea (*n*= 83). PFOS was found in the livers of 95% of the birds analyzed at concentrations greater than the limit of quantitation (LOQ=10 ng/g, w.w.). The greatest concentration of PFOS of 650 ng/g, w.w., was found in the liver of a common cormorant from the Sagami River in Kanagawa Prefecture.

Borghesi and colleagues [24] evidenced a PBDEs concentrations in Antarctic fish species ranging within average of 0.09 ng/g (w.w.) recorded in *G. nicholsi* to average of 0.44 ng/g (w.w.) measured in *C. gunnari*. In Mediterranean tuna PBDEs levels were two or three orders of magnitude higher (15 ng/g w.w.). Furthermore, PBDE congener profiles differ between species; low brominated congeners prevailed in Antarctic species while in tuna tetra- and penta-bromodiphenyl ethers are the most abundant groups (41% and 44%, respectively). In the same study, a strong correlation with the fish length is observed for the species *C. hamatus* but the same relation is not recorded considering the weight. Tuna evidences a gender dependency in PBDEs concentrations in fact levels are significantly high in females than in males (18 ng/g *vs* 13 ng/g w.w.) which authors attribute to the lower fat content in males.

4. Human exposure

PBDEs levels in humans have increased over the past several decades. Schiavone and colleagues [74] measured PBDEs in human lipid tissues from Italy even if at values lower than other POPs (PCBs and DDTs). These chemicals are structurally similar to thyroid hormones (i.e. thyroxine T4) and could disrupt thyroid homeostasis as observed in laboratory experiments on animals [75] causing damages similar to thyroid hormone deficiencies [76; 77].

Effects on male reproductive system have been documented by the literature due to the weak estrogenic/antiestrogenic activity of these chemicals [78]. In rats, the exposure to a single dose of 60 μg/kg body weight (b.w.) of PeBDE-99 produces significant decreases of sperm numbers. Akutsu and colleagues [79] evidenced relationship between human serum PBDEs and sperm quality. In particular, PBDE levels in Japan men are comparable to those found in European countries and a strong inverse correlations were observed between the serum concentration PeBDE-99 and sperm concentration (r = -0.841, p = 0.002) and testis size (r = -0.764, p = 0.01).

The Department of the Environment and Water Resources of Australia founded in 2004 a research aimed to evaluate PBDEs levels in indoor environments collecting and analysing samples from air, dust and surfaces from homes and offices in south-east Queensland. Concentrations of PBDEs were greater in indoor air than in outdoor once, evidencing that major risks are related to the indoor exposure. Furthermore, the lowest PBDE concentration in indoor dust was found in a house with no carpet, no air-conditioning, and which was older than five years. The highest concentration was found in an office with carpet and air-conditioning, and which had been refurbished in the last two years. A recent study developed by Meeker et al. [77] evidenced altered serum hormone levels in US men affected by infertility clinic (n=24) as a result of indoor exposure to PBDE. BDE 47 and 99 were detected in 100% of dust samples, and BDE 100 was detected in 67% of dust samples. A significant inverse relationship between dust PBDE concentrations and free androgen index was observed. Furthermore, dust PBDE concentrations were inversely associated with luteinizing hormone (LH) and follicle stimulating hormone (FSH), and positively associated with inhibin B and sex hormone binding globulin (SHBG).

Concerning POPs levels in humans, infants show the higher feeding exposure compared to adults due to their high feed consumption per kilogram of body weight. Weijs *et al.*, [80] evidenced in not-breastfed Dutch infants a progressively increasing exposure to POPs during the first year growth from the birth due to the diet changes. Concerning PBDEs, the mean level measured in breast milk was 3.93±1.74 ng/g lipid and the estimated PBDE daily intake for a breastfed infant was 20.6 ng/kg b.w./day after delivery [81].

Chao and colleagues [82] evidenced that, in Taiwan, PBDEs levels in breast milk (n= 46) are associated with demographic parameters, socioeconomic status, lifestyle factors, and occupational exposure. Average levels measured in 2010 (1.07-3.59 ng/g lipid) were 0.7-fold lower than in 2000. Furthermore higher levels of PBDEs were positively correlated to the maternal age and are not correlated with maternal pre-pregnant BMI (Body mass index), parity, and lipid contents of breast milk.

PBDEs level in breast milk is lower in more educated women after controlling for age and pre-pregnancy BMI in tested mothers, nevertheless these results are not completely in agreement with Wang and colleagues [83] which evidenced that the mean level of BDE47 in breast milk from mothers with pre-pregnant BMI <22.0 kg/m^2 had a significantly higher magnitude compared to those with pre-pregnant BMI >22.0 kg/m^2 (1.59 *vs* 0.995 ng/g lipid, p= 0.041) and no relationships between PBDEs exposure levels and women's age, parity, blood pressure, annual household income, and education level.

Evidences regarding a relationship between PBDEs levels in breast milk and seafood consumptions in Taiwan has been explored by the literature [83]. Women eating more fish and meat show not significantly higher PBDE levels than others nevertheless, a significant difference in PBDE levels was demonstrated between the higher (2.15 ng/g lipid) and lower (3.98 ng/g lipid) shellfish consuming subjects (p = 0.002) after an adjustment for the confounders.

Concerning ratios (PCB153/BDE47, PCB153/BDE153, PCB153/PBDEs) a significant correlation with frequent consumption of fish and shellfish is observed.

5. Ecotoxicological effects

Data collected by 3M [84; 85] on ecotoxicological effects of PFOS on aquatic species evidenced after 96h of exposure an $E_bC_{50(biomass)}$ of 71 mg/L, an $E_gC_{50(growth\ rate)}$ of 126 mg/L, and a $NOEC_{(biomass/growth\ rate)}$ of 48 mg/L on *Selenastrum capricornutum* (Algae). Acute effects on the oyster shell deposition are observed at 2.1 mg/L (96h NOEC) whereas subchronic/chronic effects were recorded on *Mysidopsis bahia* (mysid. shrimp) at 0.25 mg/L (35-day NOEC reproduction/growth). The same species evidenced respectively acute effects at 4.0 mg/L (96h EC_{50}) and 1.2 mg/L (96h NOEC).

The species *Crassostrea virginica* evidenced acute toxicity at levels higher than 3.0 mg/L (96h EC_{50}) and 96h NOEC at 1.9 mg/L.

Acute toxicity is observed in *Daphnia magna* exposing animals at 66 mg/L (48h EC_{50}) whereas chronic effects are observed over 7 mg/L (28day $NOEC_{reproduction}$).

Unio compalmatus (freshwater mussel) evidenced 96h LC_{50} of 59 mg/L and 96h NOEC of 20 mg/L.

Sanderson and colleagues [86] evidenced the ecological effects induced by the exposure to perfluorinated surfactants (PFOS and PFOA), on zooplankton species performing using classical ecotoxicological tests, 30-L indoor microcosm and 12,000-L outdoor microcosm experiments. The zooplankton community considered in this experiment was composed by the following representative species: *Cyclops diaptomus, C. strenuus, Canthocamptus staphylinus, D. magna, Keratella quadrata, Phyllopoda sp., Echninorhynchus sp., Ostracoda sp.*, and total *Rotifera sp*. In addition to zooplankton and pond snails, occasional macrophytes (*Elodea canadansis* and *Myriophyllum spicatum*), and larger invertebrates (*Ephemeroptera sp., Assellus aquaticus*) were present.

Results evidenced that zooplankton had lower tolerance toward PFOS than toward PFOA. Researchers observed that "with increasing concentrations the zooplankton community became simplified toward more robust rotifer species, which, as an indirect effect, increased their abundance due to a shift in competition and predation".

Concerning PFOA, classical ecotoxicological tests results on $LOEC_{community}$ are not available, whereas 30-L indoor and 12,000-L outdoor produce similar results ($LOEC_{community}$ 30-70 mg/L).

Concerning PFOS, classical ecotoxicological tests evidenced $LOEC_{community}$ ranges within 13-50 mg/L, while 30-L indoor and 12,000-L outdoor exposure tests produce a $LOEC_{community}$ respectively of 1-10 mg/L and 10-30 mg/L.

Concerning fishes the species *Pimephales promelas* (fathead minnow) shows 96h LC_{50} of 10 mg/L and 96h NOEC of 3.6 mg/L, the species *Lepomis macrochirus* (bluegill sunfish) has a 96h LC_{50} of 7.8 mg/L and 96h NOEC of 4.5 mg/L.

Chronic exposure of the fathead minnow *Pimephales promelas* reported 42-day $NOEC_{survival}$ of 0.33 mg/L and 47-day early life LOEC of 0.65 mg/L.

Functional studies evidenced that PFOS inhibits gap junction intercellular communication (GJIC) in rat liver epithelial cells cultured *in vitro* (personal comunication reported in [29]) and that it is an uncoupler of phosphorylation in rat liver mitochondria (personal comunication reported in [29]).

PBDEs can inhibit growth in colonies of plankton and algae and depress the reproduction of zooplankton.

Laboratory mice and rats have also shown liver function disturbances and damage to developing nervous systems as a result of exposure to PBDEs (http://www.environment.gov.au/settlements/chemicals/bfrs/index.html).

Ecotoxicological tests performed on PBDEs on different species and medium following exposed expressing results as: $LC_{50(\text{median lethal dose})}$, $LOAEL_{(\text{Lowest-Observed-Adverse-Effect Level})}$, $LOEC_{(\text{Lowest-Observed-Effect Concentration})}$, $NOAEL_{(\text{No-Observed-Adverse-Effect Level})}$, and $NOEC_{(\text{No-Observed-Effect Concentration})}$. The use of the letter "a" following data means that in the study reported highest concentration (or dose) tested did not result in statistically significant results. Since the NOEC or NOAEL could be higher, the NOEC or NOAEL are described as being greater than or equal to the highest concentration (or dose) tested.

As reported by CMABFRIP [87], *Daphnia magna* (younger than 24h old at the start of the exposure) exposed to a PeBDE mixture containing 33.7% of tetra-BDE, 54.6% of penta-BDE, and 11.7% hexa-BDE following the GLP, protocol based on OECD (Organisation for Economic Co-operation and Development) 202, TSCA (*Toxic Substances Control Act)* Title 40 and ASTM E1193-87, evidences reported levels:

- 17 μg/L (96-hour $EC_{50\text{mortality/immobility}}$),
- 20 μg/L (21-day $LOEC_{\text{mortality/immobility}}$),
- 9.8 μg/L (21-day $NOEC_{\text{mortality/immobility}}$),
- 14 μg/L (7- to 21-day $EC_{50\text{mortality/immobility}}$),
- 14 μg/L (21-day $EC_{50\text{reproduction}}$),
- 9.8 μg/L (21-day $LOEC_{\text{growth}}$),
- 5.3 μg/L (21-day $NOEC_{\text{growth}}$),
- 9.8 μg/L $LOEC_{(\text{overall study})}$,
- 5.3 μg/L $NOEC_{(\text{overall study})}$.

Exposing *Daphnia magna* to an OBDE mixture composed by 5.5% hexa-BDE, 42.3% hepta-BDE, 36.1% octa-BDE, 13.9% nona-BDE, 2.1% deca-BDE (European Communities 2003) for

21 days with the same protocol adopted for the other exposure (GLP, protocol based on OECD 202, ASTM E1193-87 and TSCA Title 40), results were the follow [13]:

- 21-day $LOEC_{(survival, reproduction, growth)}$>2.0 μg/$L_{(nominal)}$ or 1.7 μg/$L_{(measured)}$
- 21-day $NOEC_{(survival, reproduction, growth)}$>=2.0 μg/$L_{(nominal)}$ or 1.7 μg/$L_{(measured)}$[a]
- 21-day $EC_{50(survival, reproduction, growth)}$> 2.0 μg/$L_{(nominal)}$ or 1.7 μg/$L_{(measured)}$

The Great Lakes Chemical Corporation [88], reported ecotoxicological results obtained on adults of the species *Lumbriculus variegatus. The exposure mixture of* PeBDE is composed by the 0.23% tri-BDE, 36.02% of tetra-BDE, 55.10% penta-BDE, 8.58% hexa-BDE and the exposure protocol is GLP, protocol based on Phipps et al. [89], ASTM E1706-95b and U.S. EPA OPPTS (Office of Prevention, Pesticides and Toxic Substances) No. 850.1735. Animals were exposed at 23 ± 2°C, pH 7.9-8.6, DO 6.0-8.2 mg/L, hardness 130 mg/L as $CaCO_3$. On artificial sediment with the following characteristics: pH 6.6, water holding capacity 11%, mean organic matter <2%, 83% sand, 11% clay, 6% silt.

Results collected were the follows:

- 28-day LOEC $_{(survival/reproduction)}$ = 6.3 mg/kg dw of sediment
- 28-day $NOEC_{(survival/reproduction)}$ = 3.1 mg/kg dw of sediment
- 28-day $EC_{50(survival/reproduction)}$ > 50 mg/kg dw of sediment
- $growth_{(dry\ weights)}$ not significantly different from solvent control and not concentration-dependent.

The exposure of the same species to an OBDE (DE-79) mixture characterized by the 78.6% bromine content following the GLP, protocol based on Phipps et al. [89], ASTM E1706-95b and U.S. EPA OPPTS 850.1735, evidenced the following results [90; 91]:

- 28-day $LOEC_{(survival/reproduction, growth)}$> 1,340 (2% Organic carbon, OC) or 1272 (5% OC) mg/kg dw of sediment,
- 28-day $NOEC_{(survival/reproduction, growth)}$>= 1,340 (2% OC) or 1,272 (5% OC) mg/kg dw of sediment,
- 28-day $EC_{50(survival/reproduction, growth)}$> 1,340 (2% OC) or 1,272 (5% OC) mg/kg dw of sediment,
- For 2% OC study: average individual dry weights for treatments statistically lower than in control; not considered treatment-related by authors, as average biomass in treatments comparable to control.

The exposure of the adult earthworm *Eisenia fetida to an OBDE (DE-79) mixture at* 78.6% bromine content following the GLP, protocol based on U.S. EPA OPPTS 850.6200, OECD 207 and proposed OECD (2000) guideline on artificial soil (sandy loam, 69% sand, 18% silt, 13% clay, 8.0% organic matter, 4.7% carbon) at 17-21°C with a photoperiod of 16:8 light:dark, pH 5.9-6.8, soil moisture 22.0-33.5%.

Results [92] obtained are the follow:

- 28-day $LOEC_{(mortality)}$ > 1,470 mg/kg dry soil
- 28-day $NOEC_{(mortality)}$ >= 1,470 mg/kg dry soil[a]
- 28-day EC_{10}, $EC_{50(survival)}$ > 1,470 mg/kg dry soil
- 56-day $LOEC_{(reproduction)}$> 1,470 mg/kg dry soil
- 56-day $NOEC_{(reproduction)}$ >= 1,470 mg/kg dry soil[a]
- 56-day EC_{10}, $EC_{50(reproduction)}$ > 1,470 mg/kg dry soil

ACCBFRIP [93] reported for the exposure of the species *Eisenia fetida to a* DBDE mixture composed by the 97.90% of deca-BDE, the following results:

- 28-day $LOEC_{(survival)}$> 4,910 mg/kg dry $soil_{(mean\ measured)}$
- 28-day $NOEC_{(survival)}$ >= 4,910 mg/kg dry soil $_{(mean\ measured)a}$
- 28-day EC_{10}, $EC_{50(survival)}$ > 4,910 mg/kg dry soil $_{(mean\ measured)}$
- 56-day $LOEC_{(reproduction)}$> 4,910 mg/kg dry soil $_{(mean\ measured)}$
- 56-day $NOEC_{(reproduction)}$ >= 4,910 mg/kg dry soil $_{(mean\ measured)a}$
- 56-day EC_{10}, $EC_{50(reproduction)}$ > 4,910 mg/kg dry soil $_{(mean\ measured)}$

On the contrary, the exposure of *Lumbriculus variegatus to a* DBDE mixture composed by 97.3% of deca-BDE and 2.7% of other (not specified) composite from three manufacturers), evidenced the following results [94; 95]:

- 28-day $NOEC_{(survival/reproduction,\ growth)}$>= 4,536 (2.4% OC) or 3,841 (5.9% OC) mg/kg dw of sediment,
- 28-day $LOEC_{(survival/reproduction,\ growth)}$ > 4,536 (2.4% OC) or 3,841 (5.9% OC) mg/kg dw of sediment,
- 28-day $EC_{50(survival/reproduction,\ growth)}$ > 4,536 (2.4% OC) or 3,841 (5.9% OC) mg/kg dw of sediment.

The Great Lakes Chemical Corporation [96], reported ecotoxicological results obtained on the species *Zea mays* corn. *The exposure mixture was the same adopted on the earthworm* [97] and the exposure protocol is GLP, protocol based on U.S. EPA OPPTS Nos. 850.4100 and 850.4225 and OECD 208 (based on 1998 proposed revision). Plants were exposed on artificial soil (92% sand, 8%clay and 0% silt), with pH 7.5, organic matter content 2.9% and watering with well water using subirrigation (14:10 light:dark photoperiod, 16.0-39.9°C, relative humidity 19-85%).

Results collected were the follows:

- no apparent treatment-related effects on seedling emergence,
- 21-day LC_{25}, $LC_{50\ (seedling\ emergence)}$ > 1,000 mg/kg soil dw,

- mean shoot height significantly reduced at 250, 500 and ,1000 mg/kg soil dw relative to controls,
- 21-day EC_{25}, $EC_{50\ (mean\ shoot\ height)}$ > 1,000 mg/kg soil dw,
- mean shoot weight significantly reduced at 62.5, 125, 250, 500 and 1,000 mg/kg soil dw relative to controls,
- 21-day $EC_{25\ (mean\ shoot\ weight)}$ = 154 mg/kg soil dw,
- 21-day $EC_{50\ (mean\ shoot\ weight)}$ > 1,000 mg/kg soil dw,
- 21-day LOEC $_{(mean\ shoot\ weight)}$ = 62.5 mg/kg soil dw
- 21-day EC_{05} and $_{(estimated)}$ NOEC $_{(mean\ shoot\ weight)}$ = 16.0 mg/kg soil dw.

The Great Lakes Chemical Corporation [98], reported ecotoxicological results obtained on the Rat. The exposure mixture PeBDE (DE-71) was composed by 45-58.1% of penta-BDE, 24.6-35% tetra-BDE [99; 100]. Exposure doses of PeBDE were 0, 2, 10 and 100 mg/kg bw per day (doses adjusted weekly based on mean body weight of animals) and after 90 days observed effects were:

- decreased food consumption and body weight, increased cholesterol, increased liver and urine porphyrins at 100 mg/kg bw dose,
- increased absolute and relative liver weights at 10 and 100 mg/kg bw, with return to normal ranges after 24-week recovery period,
- compound-related microscopic changes to thyroid and liver at all dosage levels,
- microscopic thyroid changes reversible after 24 weeks,
- microscopic liver changes still evident at all dosage levels after 24-week recovery period,
- liver cell degeneration and necrosis evident in females at all dosage levels after 24-week recovery,
- LOAEL (liver cell damage) = 2 mg/kg bw,
- NOAEL could not be determined, as a significant effect was observed at the lowest dose tested.

Exposing rats to a DBDE (Dow-FR-300-BA) mixture with the relative composition of 77.4% of deca-BDE, 21.8% nona-BDE and 0.8% octa-BDE, results evidenced [101]:

- LOAEL $_{(enlarged\ liver,\ thyroid\ hyperplasia)}$ = 80 mg/kg bw per day,
- NOAEL = 8 mg/kg bw per day.

Little data have been gathered on the associations between PBDEs exposure and birth outcome and female menstruation characteristics in both epidemiological and animal studies. In rats, the *in utero* exposure to PBDEs reduces the number of ovarian follicles in rat females and causes permanent effects on rat males [81].

Breslin and colleagues [102] exposed rabbits to an OBDE (Saytex 111) mixture 0.2% penta-BDE, 8.6% hexa-BDE, 45.0% hepta-BDE, 33.5% octa-BDE, 11.2% nona-BDE, 1.4% deca-BDE evidencing the following results:

- no evidence of teratogenicity,
- $LOAEL_{(maternal, increased liver weight, decreased body weight gain)}$ = 15 mg/kg bw per day,
- $NOAEL_{(maternal)}$ = 5.0 mg/kg bw per day,
- $LOAEL_{(fetal, delayed ossification of sternebrae)}$ = 15 mg/kg bw per day,
- $NOAEL_{(fetal)}$ = 5.0 mg/kg bw per day.

6. Conclusions

Recent data collected on these chemicals evidence significant levels in environments, wildlife and humans. In particular, observed ecotoxicological effects on species, measured values in humans tissues and their relationships with fertility suggest that PFCs and PBDEs represent an important problem to be quickly solved. Unfortunately, collected data both on chemical distribution in abiotic and biotic matrices are fragmentary and incomplete as well as ecotoxicological studies on laboratory species and microcosms. Studies in aquatic ecosystems and, in particular, in transitional ones have to be improved to allow a correct evaluation of the exposure risk for humans to these compounds due to the dietary intakes.

Author details

Monia Renzi*

Address all correspondence to: renzi2@unisi.it

Department of Environmental Science, Via Mattioli, University of Siena, Italy

References

[1] Carson R. [1st. Pub. Houghton Mifflin, 1962]. Silent Spring. Mariner Books. 2002. ISBN 0-618-24906-0.

[2] de Azevedo e Silva CE, Azeredo A, Lailson-Brito J, Machado-Torres JP, Malm O. Polychlorinated biphenyls and DDT in swordfish (Xiphias gladius) and blue shark (*Prionace glauca*) from Brazilian coast. Chemosphere 2007;67: S48–S53.

[3] Focardi S, Bargagli R, Corsolini S. Isomer-specific analysis and toxic potential evaluation of polychlorinated biphenyls in Antarctic fish, seabirds and Weddell seals from Terra Nova Bay (Ross Sea). Antarctic Science 1995;7: 31–35.

[4] Bouloubassi I, Mejanelle L, Pete R, Fillaux J, Lorre A, Point V. PAH transport by sinking particles in the open Mediterranean Sea: A 1 year sediment trap study. Marine Pollution Bulletin 2006;52: 560–571.

[5] Corsolini S, Focardi S, Kannan K, Tanabe S, Borrell A, Tatsukawa R. Congener profile and toxicity assessment of polychlorinated biphenyls in dolphins, sharks and tuna collected from Italian coastal waters. Marine Environmental Research 1995;40: 33–53.

[6] Corsolini S, Kannan K, Imagawa T, Focardi S, Giesy JP. Polychloronaphthalenes and other dioxin-like compounds in arctic and antarctic marine food webs. Environmental Science and Technology 2002;36: 3490–3496.

[7] Storelli MM, Marcotrigiano GO. Occurrence and accumulation of organochlorine contaminants in swordfish from Mediterranean Sea: A case study. Chemosphere 2006;62: 375–380.

[8] Storelli MM, Casalino E, Barone G, Marcotrigiano GO. Persistent organic pollutants (PCBs and DDTs) in small size specimens of bluefin tuna (*Thunnus thynnus*) from the Mediterranean Sea (Ionian Sea). Environment International 2008;34: 509–513.

[9] Liem AKD. Dioxins and dioxin-like PCBs in foodstuffs. Levels and trends. Organohalogen Compounds 1999;44: 1–4.

[10] Sweetman J., Alcock R.E., Wittisiepe J., Jones K.C. 2000. Human exposure to PCDD/Fs in the UK: The development of a modeling approach to give historical and future perspectives. Environ. Int. 26: 37–47.

[11] Falandysz J, Wyrzykowskay B, Puzyny T, Strandbergy L, Rappe C. Polychlorinated biphenyls (PCBs) and their congener specific accumulation in edible fish from the Gulf of Gdansk, Baltic Sea. Food Additives and Contamination 2002;19: 779–795.

[12] Brambilla G, Cherubini G, de Filippis S, Magliuolo M, Di Domenico A. Review of aspects pertaining to food contamination by polychlorinated dibenzodioxins, dibenzofurans, and biphenyls at the farm level. Anal. Chim. Acta 2004;514(1): 1–7.

[13] CMABFRIP (Chemical Manufacturers Association Brominated Flame Retardant Industry Panel). 1997. Octabromodiphenyl oxide (OBDPO): A flow-through life-cycle toxicity test with the cladoceran (Daphnia magna). Final report. Project No. 439A-104, Wildlife International, Ltd., May.

[14] Bommanna GL, Kannan K. Global Organochiorine Contamination Trends: An Overview. Ambio 1994;23(3): 187-191.

[15] UNECE 1998. The 1998 Aarhus Protocol on Heavy Metals. On-line available at: http://www.unece.org/env/lrtap/hm_h1.html.

[16] Hargrave B, Vass W, Erickson P, Fowler B. Atmospheric transport of organochlorines to the arctic ocean. Tellus 1988;40(B): 480–493.

[17] Fernandez P, Grimalt JO. On the global Distribution of Persistent Organic Pollutants. Chimia 2003;57: 514-521.

[18] Richardson SD, Ternes TA. Water analysis: emerging contaminants and current issues. Analytical Chemistry 2005;77: 3807-3838.

[19] Van de Vijver KI, Hoff P, Das K, Brasseur S, Dongen WV, Esmans E, Reijnders P, Blust R, Coen WD. Tissue Distribution of Perfluorinated Chemicals in Harbor Seals (*Phoca vitulina*) from the Dutch Wadden Sea. Environmental Science and Technology 2005;39(18): 6978–6984.

[20] Yeung LWY, So MK, Jiang G, Taniyasu S, Yamashita N, Song M, Wu Y, Li J, Giesy JP, Guruge KS, Lam PKS. Perfluorooctanesulfonate and related fluorochemicals in human blood samples from China. Environmental Science and Technology 2006;40(3): 715–720.

[21] Midasch O, Schettgen T, Angerer J. Pilot study on the perfluorooctanesulfonate and perfluorooctanoate exposure of the German general population. International Journal of hygiene and environmental health 2006;209(6): 489-496.

[22] Simcik MF. Air monitoring of persistent organic pollutants in the Great Lakes: IADN vs. AEOLOS. Environmental Monitoring and Assessment 2005;100(1-3): 201-216.

[23] Prevedouros K, Cousins IT, Buck RC, Korzeniowski SH. Sources, fate and transport of perfluorocarboxylates. Environmental Science and Technology 2006;40: 32–44.

[24] Borghesi N, Corsolini S, Leonards P, Brandsma S, de Boer J, Focardi S. Polybrominated diphenyl ether contamination levels in fish from the Antarctic and the Mediterranean Sea. Chemosphere 2009;77: 693–698.

[25] Kannan K, Corsolini S, Falandysz J, Oehme G, Focardi S, Giesy J. Perfluorooctanesulfonate and related fluorinated hydrocarbons in Marine mammals, Fishes, and Birds from Coasts of the Baltic and the Mediterranean Seas. Environmental Science and Technology 2002;36: 3210-3216.

[26] Key BD, Howell RD, Criddle CS. Fluorinated organics in the biosphere. Environmental Science and Technology 1997;31: 2445-2454.

[27] Renner R. Growing concern over perfluorinated chemicals. Environmental Science and Technology 2001;35: 154A–160A.

[28] Butenhoff JL, Gaylor DW, Moore JA, Olsen GW, Rodricks J, Mandel JH, Zobel LR. Characterization of risk for general population exposure to perfluorooctanoate. Regulatory Toxicology and Pharmacology 2004;39: 363–380.

[29] Key BD, Howell RD, Criddle C. Defluorination of Organofluorine Sulfur Compounds by *Pseudomonas* Sp. Strain D2. Environmental Science and Technology 1998;32: 2283-2287.

[30] MSDS. Material Safety Data Sheet for FC-203FC Light Water Brand Aqueous Film Forming Foam, 3M Co. 1999. London, ON, Canada.

[31] Moody CA, Field JA. Perfluorinated surfactants and the environmental implications of their use in fire-fighting foams. Environmental Science and Technology 2000;34: 3864-3870.

[32] Moody CA, Martin JW, Kwan WC, Muir DG, Mabury SA. Monitoring Perfluorinated Surfactants in biota and surface water samples following an accidental release of fire-fighting foam into Etobicoke Creek. Environmental Science and Technology 2002;36: 545-551.

[33] Brooke D, Footitt A, Nwaogu TA. Environmental risk evaluation report: Perfluorooctanesulphonate (PFOS). 2004. On-line available at URL: http://www.environment-agency.gov.uk/commondata/105385/pfos_rer_sept04_864557.pdf./

[34] Kodavanti PRS, Senthil Kumar K, Loganathan BG. Organohalogen pollutants in the environment and their effects on wildlife and human health. International Encyclopedia of Public Health 2008;4: 686.693.

[35] de Boer J, de Boer K, Boon JP. Polybrominated biphenyls and diphenylethers. In: Paasivirta J. (Ed.), The Handbook of Environmental Chemistry. 2000; 61–95. Springer, Berlin.

[36] Alaee M, Arias P, Sjodin A, Bergman A. An overwiew of commercially used brominated flame retardants, their applications, their use patterns in differents countries/ regions and possible modes of release. Enviromental International 2003;9: 683-689.

[37] BSEF. Bromine Science and Environmental Forum. 2000. On-line available at: http:// www.bsef.com/

[38] Birnbaum LS, Staskal DF. Brominated flame retardants: cause for concern?. Environmental Health Perspectives 2004;112: 9-17.

[39] Palm A, Cousins IT, Mackay D, Tysklind M, Metcalfe C, Alaee M. Assessing the environmental fate of chemicals of emerging concern: a case study of the polybrominated diphenyl ethers. Environmental Pollution 2002;117(2): 195-213.

[40] Martellini T. 2008. Studio del trasporto a lungo raggio (LRT) e del destino ambientale di composti organici persistenti (POPs). Scuola di Dottorato in Scienze Dottorato in Scienze Chimiche XXI Ciclo. Università degli Studi di Firenze. Doctoral dissertation, pp. 206.

[41] EFSA 2011. European Food Safety Authority. Scientific Opinion on Polybrominated Diphenyl Ethers (PBDEs) in Food EFSA Panel on Contaminants in the Food Chain (CONTAM). EFSA Journal. 9(5):2156.

[42] Wang Y, Jiang G, Lam PKS, Li A. Polybrominated diphenyl ether in the East Asian environment: A critical review. Environment International 2007;33(7): 963–973.

[43] Sajwan KS, Kumar KS, Kelley S, Loganathan BG. Deposition of Organochlorine Pesticides, PCBs (Aroclor 1268), and PBDEs in Selected Plant Species from a Superfund Site at Brunswick, Georgia, USA. Bulletin of Environmental Contamination and Toxicology 2009;82(4): 444-449.

[44] Senthilkumar K, Ohi E, Sajwan K, Takasuga T, Kannan K. Perfluorinated Compounds in River Water, River Sediment, Market Fish, and Wildlife Samples from Japan. Bulletin of Environmental Contamination and Toxicology 2007;79: 427–431.

[45] Gouin T, Thomas GO, Cousins I, Barber J, Mackay D, Jones KC. Air-surface exchange of polybrominated diphenyl ethers and polychlorinated biphenyls. Environmental Science and Technology 2002;36(7): 1426-1434.

[46] Strandberg B, Dodder NG, Basu I, Hites RA. Concentrations and spatial variations of polybrominated diphenyl ethers and other organohalogen compounds in Great Lakes air. Environmental Science and Technology 2001; 35: 1078-1083.

[47] Harner T, Ikonomou M, Shoeib M, Stern G, Diamond M. Passive air sampling results for polybrominated diphenyl ethers along an urban-rural transect. 2002. 4th Annual Workshop on Brominated Flame Retardants in the Environment, June 17-18, Canada Centre for Inland Waters, Burlington, Ontario. pp. 51-54.

[48] Hale RC, La Guardia MJ, Harvey E, Mainor TM. Potential role of fire retardant-treated polyurethane foam as a source of brominated diphenyl ethers to the U.S. environment. Chemosphere 2002;46: 729-735.

[49] Hale RC, Alaee M, Manchester-Neesvig JB, Stapleton HM, Ikonomou MG. Polybrominated diphenyl ether (PBDE) flame retardants in the North American environment. Environment International 2003;29: 771-779.

[50] Kumar KS, Priya M, Sajwan KS, Kõlli R, Roots O. Residues of persistent organic pollutants in Estonian soils (1964-2006) Estonian Journal of Earth Sciences 2009;58(2): 109-123.

[51] Nakata H, Kannan K, Nasu T, Cho HS, Sinclair E, Takemura A. Perfluorinated Contaminants in Sediments and Aquatic Organisms Collected from Shallow Water and Tidal Flat Areas of the Ariake Sea, Japan: Environmental Fate of Perfluorooctane Sulfonate in Aquatic Ecosystems. Environmental Science and Technology 2006;40(16): 4916–4921.

[52] IFA. Institute for Occupational Safety and Health of the German Social Accident Insurance. 2001. Subdivision "Information on Hazardous Substances". Alte Heerstr. 111 D-53757 Sankt Augustin, Germany. Available on-line at: http://gestis-en.itrust.de/nxt/gateway.dll/gestis_en/036110.xml?f=templates$fn=default.htm$3.0

[53] OECD, Draft assessment of perfluorooctane sulfonate and its salts. ENV/JM/EXCH, 8. 2002. Paris, France.

[54] Boulanger B, Vargo J, Schnoor J, Hornbuckle K. Detection of Perfluorooctane Surfactants in Great Lakes Water. Environmental Science and Technology 2004;38: 4064-4070.

[55] Hansenk J, Johnson HO, Eldrige JS, Butenhoff JL, Dick LA. Quantitative Characterization of trace levels of PFOS and PFOA in the Tennessee River. Environmental Science and Technology 2002;36: 1681-1685.

[56] Yamashita N, Kannan K, Taniyasu S, Horii Y, Petrick G, Gamo T. A global survey of perfluorinated acids in oceans. Marine Pollution Bulletin 2005;doi:10.1016/j.marpolbul.2005.04.026.

[57] Luckey FJ, Fowler B, Litten S. Establishing baseline levels of polybrominated diphenyl ethers in Lake Ontario surface waters. Unpublished manuscript dated 2002/03/01. New York State Department of Environmental Conservation, Division of Water, 50 Wolf Road, Albany, NY 12233-3508.

[58] Stapleton HM, Baker JE. Comparing the temporal trends, partitioning and biomagnification of PBDEs and PCBs in Lake Michigan. Abstract. 2001. 3rd Annual Workshop on Brominated Flame Retardants in the Environment, August 23-24, 2001, Canada Centre for Inland Waters, Burlington, Ontario.

[59] Becker AM, Gerstmann S, Frank H. Perfluorooctanoic acid and perfluorooctane sulfonate in the sediment of the Roter Main river, Bayreuth, Germany, Environmental Pollution 2008;doi:10.1016/j.envpol.2008.05.024.

[60] Kolic TM, MacPherson KA, Reiner EJ, Ho T, Kleywegt S, Dove A, Marvin C. Brominated diphenyl ether levels: a comparison of tributary sediments versus biosolid material. Organohalogen Compounds 2004;66: 3830-3835.

[61] Rayne S, Ikonomou MG, Antcliffe B. Rapidly increasing polybrominated diphenyl ether concentrations in the Columbia River system from 1992 to 2000. Environmental Science and Technology 2003;37(13): 2847-2854.

[62] Muir D, Teixeira C, Chigak M, Yang F, D'Sa I, Cannon C, Pacepavicius G, Alaee M. Current deposition and historical profiles of decabromodiphenyl ether in sediment cores. Dioxin 2003, 23rd International Symposium on Halogenated Environmental Organic Pollutants and POPs. Organohalogen Compd. 61: 77-80.

[63] Kolic TM, MacPherson KA, Reiner EJ, Ho T, Kleywegt S, Payne M, Alaee M. Investigation of brominated diphenyl ethers in various land applied materials. 2003. Abstract. 5th Annual Workshop on Brominated Flame Retardants in the Environment, August 22-23, 2003. Boston, MA.

[64] Solé M, Hambach B, Cortijo V, Huertas D, Fernandez P, Company JB. Muscular and Hepatic Pollution Biomarkers in the Fishes Phycis blennoides and Micromesistius poutassou and the Crustacean Aristeus antennatus in the Blanes Submarine Canyon (NW Mediterranean). Archives Environmental Contamination and Toxicology 2009;57: 123–132.

[65] Bodiguel X, Loizeau V, Le Guellec AM, Roupsard F, Philippon X, Mellon-Duval C. Influence of sex, maturity and reproduction on PCB and p,p'DDE concentrations and repartitions in the European hake (Merluccius merluccius, L.) from the Gulf of Lions (N.W. Mediterranean). Science of the Total Environment 2009;408: 304–311.

[66] van Leeuwen SPJ, Karrman A, van Bavel B, de Boer J, Lindström G. Struggle for quality in determination of perfluorinated contaminants in environmental and human samples. Environmental Science and Technology 2006;doi: 10.1021/es061052c.

[67] Gulkowska A, Jiang Q, Ka M, Sachi S, Taniyasu, Lam PKS, Yamashita N. Persistent Perfluorinated acids in seafood collected from two cities of China. Environmental Science and Technology 2006;40(12): 3736–3741.

[68] Tseng CL, Liu LL, Chen CM, Ding WH. Analysis of perfluorooctanesulfonate and related fluorochemicals in water and biological tissue samples by liquid chromatography–ion trap mass spectrometry. Journal of Chromatography A 2006;1105(1-2): 119–126.

[69] Furdui VI, Stock N, Whittle DM, Crozier P, Reiner E, Muir DCG, Mabury SA. Perfluoroalkyl contaminants in lake trout from the Great Lakes. ENV024 Furdui. "Fluoros" 9th International Symposium on Fluorinated Alkyl Organics in the Environment, August 2005, Toronto, Canada.

[70] Tomy GT, Budakowski WR, Halldorson T, Helm PA, Stern GA, Friesen K. Fluorinated organic compounds in an eastern arctic marine food web. Environmental Science and Technology 2004;38: 6475–6481.

[71] Martin JW, Smithwick MM, Braune BM, Hoekstra PF, Muir DCG, Mabury SA. Identification of Long-Chain Perfluorinated Acids in Biota from the Canadian Arctic. Environmental Science and Technology 2004;38(2): 373–380.

[72] Cunha I, Hoff P, Van de Vijver K, Guilhermino L, Esmans E, De Coen W. Baseline study of perfluorooctane sulfonate occurrence in mussels, *Mytilus galloprovincialis,* from north-central Portuguese estuaries. Marine Pollution Bulletin 2005;50(10): 1128-1132.

[73] Kannan K, Choi J-W, Iseki N, Senthilkumar K, Kim D, Masunaga S, Giesy JP. Concentrations of perfluorinated acids in livers of birds from Japan and Korea. Chemosphere 2002;49: 225–231.

[74] Schiavone A, Kannan K, Horii Y, Focardi S, Corsolini S. Polybrominated diphenyl ethers, polychlorinated naphthalenes and polycyclic musks in human fat from Italy: Comparison to polychlorinated biphenyls and organochlorine pesticides. Environmental Pollution 2010;158: 599–606.

[75] Zhou T, Taylor MM, DeVito MJ, Crofton KM. Developmental exposure to brominated diphenyl ethers results in thyroid hormone disruption. Toxicological Science 2002;66: 105–116.

[76] Viberg H, Fredriksson A, Eriksson P. Investigations of strain and/or gender differences in developmental neurotoxic effects of polybrominated diphenyl ethers in mice. Toxicological Science 2004;81: 344–353.

[77] Meeker JD, Johnson PI, Camann D, Hauser R. Polybrominated diphenyl ether (PBDE) concentrations in house dust are related to hormone levels in men. Science of the Total Environment 2009;407: 3425–3429.

[78] Meerts IA, Letcher RJ, Hoving S, Marsh G, Bergman A, Lemmen JG, van der Burg B, Brouwer A. In vitro estrogenicity of polybrominated diphenyl ethers, hydroxylated PBDEs, and polybrominated bisphenol A compounds. Environ Health Perspect 2001;109: 399–407.

[79] Akutsu K, Takatori S, Nozawa S, Yoshiike M, Nakazawa H, Hayakawa K, Makino T, Iwamoto T. Polybrominated Diphenyl Ethers in Human Serum and Sperm Quality. Bulletin of Environmental Contaminant and Toxicology 2008;80: 345–350.

[80] Weijs PJM, Bakker MI, Korver KR, van Goor Ghanaviztchi K, van Wijnen JH. Dioxin and dioxin-like PCB exposure of non-breastfed Dutch infants. Chemosphere 2006;64(9): 1521-1525.

[81] Chao HR, Wang SL, Lee WJ, Wang YF, Päpke O. Levels of polybrominated diphenyl ethers (PBDEs) in breast milk from central Taiwan and their relation to infant birth outcome and maternal menstruation effects. Environment International 2007;33: 239–245.

[82] Chao HA, Chen SCC, Chang CM, Koh TW, Chang-Chien GP, Ouyang E, Lin SL, Shy CG, Chen FA, Chao HR. Concentrations of polybrominated diphenyl ethers in breast milk correlated to maternal age, education level, and occupational exposure. Journal of Hazardous Materials 2010;175: 492–500.

[83] Wang YF, Wang SL, Chen FA, Chao HA, Tsou TC, Shy CG, Papke O, Kuo YM, Chao HR. Associations of polybrominated diphenyl ethers (PBDEs) in breast milk and dietary habits and demographic factors in Taiwan. Food and Chemical Toxicology 2008;46: 1925–1932.

[84] 3M 2000. Sulfonated perfluorochemicals in the environment: sources, dispersion, fate and effects. Docket submitted to the US EPA, Washington, DC.

[85] 3M 2003. Health and environmental assessment of perfluorooctane sulfonic acid and its salts. US EPA docket No. AR-226-1486. US Environmental Protection Agency, Washington, DC.

[86] Sanderson H, Boudreau TM, Mabury SA, Solomon KR. Effects of perfluorooctane sulfonate and perfluorooctanoic acid on the zooplanktonic community. Ecotoxicology and Environmental Safety 2004;58: 68-76.

[87] CMABFRIP (Chemical Manufacturers Association Brominated Flame Retardant Industry Panel). 1998. Pentabromodiphenyl oxide (PeBDPO): A flow-through life-cycle

toxicity test with the cladoceran (Daphnia magna). Project No. 439A-109, Wildlife International, Ltd., September.

[88] Great Lakes Chemical Corporation. 2000. Pentabromodiphenyl oxide (PeBDPO): A prolonged sediment toxicity test with *Lumbriculus variegatus* using spiked sediment. Project No. 298A-109, Wildlife International, Ltd., April.

[89] Phipps GL, Ankley GT, Benoit DA, Mattson VR. Use of the aquatic oligochaete *Lumbriculus variegatus* for assessing the toxicity and bioaccumulation of sediment-associated contaminants. Environmental Toxicology and Chemistry 1993;12: 269-279.

[90] Great Lakes Chemical Corporation. 2001. Octabromodiphenyl ether: A prolonged sediment toxicity test with *Lumbriculus variegatus* using spiked sediment with 2% total organic carbon. Final report. Project No. 298A-112, Wildlife International, Ltd., February.

[91] Great Lakes Chemical Corporation. 2001. Octabromodiphenyl ether: A prolonged sediment toxicity test with *Lumbriculus variegatus* using spiked sediment with 5% total organic carbon. Final report. Project No. 298A-113, Wildlife International, Ltd., February.

[92] Great Lakes Chemical Corporation. 2001. Effect of octabromodiphenyl oxide on the survival and reproduction of the earthworm, *Eisenia fetida*. Study No. 46419, ABC Laboratories, Inc., December.

[93] ACCBFRIP (American Chemistry Council Brominated Flame Retardant Industry Panel). 2001. Effect of decabromodiphenyl oxide (DBDPO) on the survival and reproduction of the earthworm, *Eisenia fetida*. Final report. Study No. 465440, ABC Laboratories, Inc., December.

[94] ACCBFRIP (American Chemistry Council Brominated Flame Retardant Industry Panel). 2001. Decabromodiphenyl ether: A prolonged sediment toxicity test with *Lumbriculus variegatus* using spiked sediment with 2% total organic carbon. Final report. Project No. 439A-113, Wildlife International, Ltd., February.

[95] ACCBFRIP (American Chemistry Council Brominated Flame Retardant Industry Panel). 2001. Decabromodiphenyl ether: A prolonged sediment toxicity test with *Lumbriculus variegatus* using spiked sediment with 5% total organic carbon. Final report. Project No. 439A-114, Wildlife International, Ltd., February.

[96] Great Lakes Chemical Corporation. 2000. Pentabromodiphenyl oxide (PeBDPO): A toxicity test to determine the effects of the test substance on seedling emergence of six species of plants. Final report. Project No. 298-102, Wildlife International, Ltd., April.

[97] Great Lakes Chemical Corporation. 2000. Analytical method verification for the determination of pentabromodiphenyl oxide (PeBDPO) in soil to support an acute toxicity study with the earthworm. Final report. Project No. 298C-117, Wildlife International, Ltd., February.

[98] Great Lakes Chemical Corporation. 1984. 90-day dietary study in rats with pentabromodiphenyl oxide (DE-71). Final report. Project No. WIL-12011,WIL Research Laboratories, Inc.

[99] Sjödin A. Occupational and dietary exposure to organohalogen substances, with special emphasis on polybrominated diphenyl ethers. 2000. Doctoral dissertation, Stockholm University.

[100] Zhou T, Ross DG, DeVito MJ, Crofton KM. Effects of short-term in vivo exposure to polybrominated diphenyl ethers on thyroid hormones and hepatic enzyme activities in weanling rats. Toxicology Science 2001;61: 76-82.

[101] Norris JM, Ehrmantraut JW, Gibbons CL, Kociba RJ, Schwetz BA, Rose JQ, Humiston CG, Jewett GL, Crummett WB, Gehring PJ, Tirsell JB, Brosier JS. Toxicological and environmental factors involved in the selection of decabromodiphenyl oxide as a fire retardant chemical. Journal of Fire Flammants Combustible Toxicology 1974;1: 52-77.

[102] Breslin WJ, Kirk HD, Zimmer MA. Teratogenic evaluation of a polybromodiphenyl oxide mixture in New Zealand White rabbits following oral exposure. Fundaments of Applied Toxicology 1989;12: 151-157.

Chapter 5

The Investigation and Assessment on Groundwater Organic Pollution

Hongqi Wang, Shuyuan Liu and Shasha Du

Additional information is available at the end of the chapter

1. Introduction

Groundwater is an important part of the water resource. It plays an irreplaceable role in supporting the national economy and social development. In China, more than 1/3 of the total water resources are utilized. As surveys shown, over 400 cities of all exploit groundwater. More seriously, many of them use groundwater as the only source of water supply.

A series of problems emerge gradually with the utilization of groundwater. Just as river waters have been over-used and polluted in many parts of the world, so have groundwater. The organic solvents and dioxins pollution of Love Canal occurred in 1978 is one of the most widely known examples, which contributes high rates of cancer and an alarming number of birth defects. Similar things occur frequently in recent decades. Governance of groundwater is so urgent a major matter of peace and prosperity. After years of researches, the nature and pollution mechanism of the contaminants in the groundwater have already got comprehended.

General scope of the organic contamination in groundwater is reviewed in this chapter. We will detail account the types of groundwater organic contamination, the pollution source of groundwater. and the fate and transport of chemicals in groundwater. Also a detailed description of the investigation and assessment method in this chapter. At last, we give some comments and suggestion on the groundwater investigation and assessment.

The figure 1 described some source of groundwater contamination, and the transport of chemicals in groundwater. We can see the landfills, leaking sewers, oil storage tanks, pesticides and fertilizer, and septic tank in the picture, all of these could be the pollution source of groundwater. We also can know the groundwater transport and flow in the unsaturated zone and saturated zone.

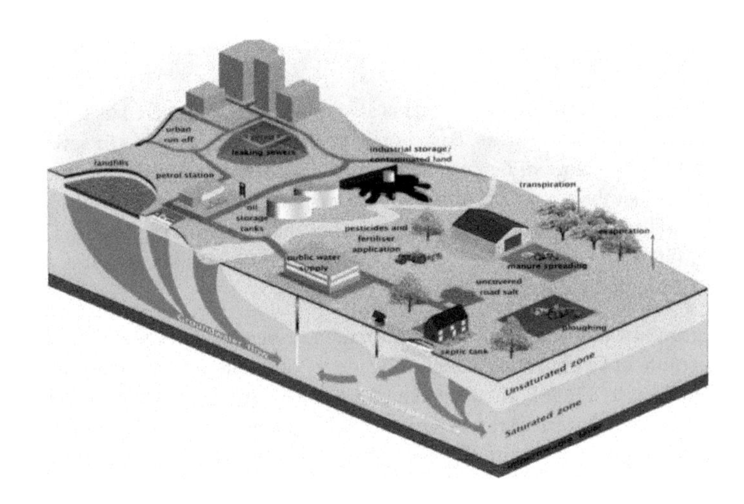

Figure 1. Diagram of the Groundwater Pollution [1].

2. A review of organic compounds in groundwater

Many studies have been conducted since 1970 to characterize concentrations of organic compounds in groundwater. In 1977, 16 drinking water wells have been closed in Gray town of Maine state because of there were at least 8 synthetic organics that were detected in drinking water wells. And in 1986, there were at least 33 organics that were detected in drinking water wells in USA [2]. It has been reported trace organic pollutants to be detected in all of 50 states. The U.S. Geological Survey(USGS) collected and assorted the test data the 1926 drinking water wells in the nation's rural areas from 1986 to 1999. And at least one VOCs were detected from 232 wells and the positive rate was 12%, with the highest positive rate were Chloroform, tetrachloroethylene and so on [3].

Similar conditions are to be found in other countries. In the 80s of last century, based on an inventory of the presence of halogenated substances in raw water of 232 groundwater pumping stations in The Netherlands a compilation of more than 100 organic substances identified in contaminated groundwater, the detection rate of trichloroethylene up to 67% [4]. The organic pollutants could be detected in groundwater in Britain. Flordward studied on 209 water supply wells in Britain shown that the main pollutant in the groundwater are the trichloroethylene and tetrachloroethylene. Beginning in 1974, Environment Agency of Japan conducted a nationwide comprehensive survey of chemical environmental safety. The trichloroethylene in groundwater was reported for the first time. The European Union is the largest pesticide consumer in the world, more than 600 pesticides were applied. Six of the top 10 were European countries in the pesticide application. Atrazine exists in groundwater all over the Europe, and the content always beyond the European Union drinking water standard(0.1μg/l) 10-100 times.

The research on organic pollutions of China is at starting stage, but there has serious organic contamination events in some areas. Based on a study of water pollution in sewage system in the Gaobeidian Prefecture of Beijing, 1988, organic substances identified in shallow wells and deep wells in South-east agriculture districts in Beijing. And 32 organic substances identified in deep wells, and 52 in shallow wells. Most of that are carcinogens (e.g. Chloroform and benzene) [5]. Analysis of the years groundwater monitoring data, and it is shown that the quality of groundwater is gradually worse.

A study on groundwater organic pollution in region of Beijing, Tianjin and Tangshan conducted by Institute of Chemistry of Chinese Academy of Sciences shows that the type of organic pollutants up to 133 [6]. The researchers Chen Honghan, He Jiangtao and others [7] have summed up the characteristics of organic contamination of shallow groundwater in a study area of the Taihu Lake basin. The results show that the detection probabilities of compounds in groundwater are higher but the concentrations of the compounds are lower. The concentrations of all the components of BTEX and halocarbons are lower than the standards set by the U.S. Environmental Protection Agency (EPA) for drinking water except for benzene in a few sampling sites.

3. Types of groundwater organic contamination

Different types of groundwater contamination sources can pose different threats to human health and different problems in health risk assessment (table 1).

3.1. Volatile organic compounds (VOCs)

Volatile organic compounds (VOCs) are organic compounds with chemical and physical properties that allow the compounds to move freely between water and air. VOCs have been used extensively in industry, commerce, and households in the United States since the 1940's. Many products contain VOCs including fuels, solvents, paints, glues, adhesives, deodorizers, refrigerants, and fumigants. In general, these compounds have low molecular weights, high vapor pressures, and low-to-medium water solubilities [8]. Many of these compounds show evidence of animal or human carcinogenicity, mutagenicity, or teratogenicity. And these compounds are quite persistent in groundwater, because of their relatively low biological and chemical reactivity. This persistence is assisted by low temperatures, absence of light and contact with the atmosphere, and comparatively low microbial concentrations typical of groundwater environments. By comparison with other organic compounds, VOCs may be transported for relatively long distances in groundwater, as a result of their relatively weak sorption affinity and their resistance to degradation. Because of human-health concerns, many VOCs have been the focus of national regulations, monitoring, and research during the past 10 to 20 years.

3.2. The pesticides

Pesticides consist of a large group of chemicals that are used in agriculture and residential settings to control plant and animal infestation. Pesticides are commonly applied on farms, fruit orchards, golf courses, and residential lawns and gardens. There are several different types of pesticides: Herbicides, Insecticides, Nematocides, Fungicides. Some pesticides do not break down easily in water and can remain in the groundwater for a long period of time. Likewise, the insecticide DDT, though banned for nearly twenty years, can still be found at trace levels in some groundwater. After prolonged exposure to high doses, some pesticides can cause cancer; some can also result in birth defects and damage to the nervous system. The use of pesticides and herbicides is one of the main ways of organic pollution of groundwater. Many water wells and irrigation wells have been closed for the byproducts from pesticides and herbicides be detected in shallow water in Colorado.

Ordering	Component	CASRN	Types
1	Trichloromethane	67-66-3	VOCs
2	Tetrachloroethylene	127-18-4	VOCs
3	1,1,1-Trichloroethane	71-55-6	VOCs
4	Trichloroethylene	79-01-6	VOCs
5	1,1-Dichloroethene	75-35-4	VOCs
6	Methyl tert-butyl ether	1634-04-4	VOCs
6*	cis-1,2-Dichloroethylene	156-59-2	VOCs
8	1, 2, 4-Trimethylbenzene	95-63-6	VOCs
9	Toluene	108-88-3	VOCs
10	Prometon	1610-18-0	Pesticides
11	1,1-Dichloroethane	75-34-3	VOCs
12	Bromacil	314-40-9	Pesticides
13	Tebuthiuron	34014-18-1	Pesticides
14	1, 3-Dichlorobenzene	541-73-1	VOCs
15	1,2-Dichloropropane	78-87-5	VOCs
16	Carbon disulfide	75-15-0	VOCs
17	Deethylatrazine	6190-65-4	Pesticides
17*	1,4-Dichlorobenzene	106-46-7	VOCs
19	Sulfamethoxazole	723-46-6	Medicine
20*	1,2-Dichlorobenzene	95-50-1	VOCs
20	2-Hydroxyatrazine	2163-68-0	Pesticides

22*	Trichlorofluoromethane	75-69-4	VOCs
22	Bentazon	25057-89-0	Pesticides
24	Atrazine	1912-24-9	Pesticides
25	Picloram	1918-2-1	Pesticides
26	Diuron	330-54-1	Pesticides
27*	Benzene	71-43-2	VOCs
27*	Tetrachloromethane	56-23-5	VOCs
29	Chlorobenzene	108-90-7	VOCs
30*	2-Butanone	78-93-3	VOCs
30	Acetone	67-64-1	VOCs
32*	m- + p-Xylene	106-42-3	VOCs
32*	trans-1,2-Dichloro- ethylen	156-60-5	VOCs
32*	1,2-Dibromoethane	106-93-4	VOCs
35	Ethylbenzene	100-41-4	VOCs
36	caffeine	58-08-2	Medicine
37	Isopropylbenzene	98-82-8	VOCs
38	o-Xylene	95-47-6	VOCs
38*	1,1,2-Trichloroethane	79-00-5	VOCs
38*	Bromodichloromethane	75-27-4	VOCs
38*	1,1,1,2-Tetrachloroethane	630-20-6	VOCs
38*	n-Propylbenzene	103-65-1	VOCs
43	Chloromethane	74-87-3	VOCs
44	1,1,2-Trichloro-1,2,2-trifluoroethane	76-13-1	VOCs
45	Dichlorodifluoromethane	75-71-8	VOCs
46	Metolachlor	51218-45-2	Pesticides
46*	Simazine	122-34-9	Pesticides
48	Bromoform	75-25-2	VOCs
48*	Imidacloprid	138261-41-3	Pesticides
48*	1,3,5-Trimethylbenzene	108-67-8	VOCs

*show the same detection with the front component. CASRN is the register number of chemical substances formulate by Chemical Abstracts Service, m means meta-position, p means para-position.

Table 1. 50 organic pollutants most commonly detected in groundwater [11]

3.3. The other organic contamination

Tens of thousands of manmade chemicals are used in today's society with all having the potential to enter our water resources. There are a variety of pathways by which these organic contaminants can make their way into the aquatic environment [9]. If the groundwater is the drinking water sources, there will be potentially dangerous on human health. Pharmaceuticals and other organic contaminants are a set of compounds that are receiving an increasing amount of public and scientific attention. Water samples were collected from a network of 47 groundwater sites across 18 states in 2000 [10]. All samples collected were analyzed for 65 organic contaminants representing a wide variety of uses and origins. Thus, sites sampled were not necessarily used as a source of drinking water but provide a variety of geohydrologic environments with potential sources of organic contaminants. organic contaminants were detected in 81% of the sites sampled, with 35 of the 65 organic contaminants being found at least once. The most frequently detected compounds include N,N-diethyltoluamide (35%, insect repellant), bisphenol A (30%, plasticizer), tri(2-chloroethyl) phosphate (30%, fire retardant), sulfamethoxazole (23%, veterinary and human antibiotic), and 4-octylphenol monoethoxylate (19%, detergent metabolite).

4. Sources of groundwater contamination

Organic contamination includes all of natural and synthetic that could cause adverse effect on human health or ecology environment.

4.1. Natural pollution sources

Naturally formed waters such as ocean water and connate brines can be sources of groundwater contamination under certain circumstances. Changes in pumping rates can cause fresh-water aquifers to be contaminated by intrusion of seawater. Similarly, changes in the groundwater flow field or leakage through imperfectly sealed wells can cause contamination of groundwater supply by naturally occurring brines or other poor-quality waters. Generally, trace amount of natural organic compounds existence in groundwater in most of regions. The major is humic acid, especially in forest and grassland. Although itself could not impair the groundwater quality, it could be enhance the heavy metal and other organic matters activities in groundwater.

4.2. Organic contamination come from human activities

As the human population grows, groundwater pollution from human activity also increases. There are a number of possible sources that could lead to groundwater contamination. Such as crude oil leakage in oil production, organic waste discharge, spills and leaks from underground storage tank and so on.

4.2.1. City and industrial wastewater

The treatment and disposal of sewage present health risks in both developed and undeveloped countries. In undeveloped countries, sewage may be directly applied to the land surface. In more developed areas, sewage is generally transported to municipal treatment plants or disposed of in septic tanks and cesspools. Groundwater contamination can result in all these cases. Sewage provides a source of pathogens, nitrates, and a variety of organic chemicals to groundwater. Land application of sewage can provide a direct contaminant source via infiltration. Treatment plants can act as contaminant sources in several ways. Leaks may occur in sewer lines and infiltration may occur from the ponds and lagoons within the treatment plants. In addition, the sewage sludge that is a product of sewage treatment processes is often disposed on land in conjunction with agricultural activity. Depending on the characteristics of the sludge, the soil characteristics, and the application process, such land application can act as a large non-point source of groundwater contamination. Land disposal of treated waste water can pose comparable risks. Depending on hydrogeologic conditions, septic tanks and cesspools may allow untreated sewage to enter the groundwater flow system. In addition, use of solvents to clean out the systems can cause groundwater contamination by synthetic organic compounds. The material cleaned out from septic tanks must eventually be disposed of, often by land application.

Industrial Wastewaters are applied to land in ponds or lagoons that are either designed to percolate the liquid into the soil or to store and/or evaporate the liquid above ground. In either case, such facilities act as potential groundwater contamination sources. Facilities designed to intentionally infiltrate into the ground include cooling ponds for power generation and for other industrial processes. The liquids in such facilities may contain potentially hazardous materials. Storage and evaporation ponds are often lined to prevent infiltration, but are likely to act as groundwater contamination sources under some circumstances, depending on surface runoff characteristics, the integrity and permeable of the liner(s), and the groundwater flow system. Poorly designed evaporation ponds may, in many cases, function as infiltration ponds.

In the United States, the big city and small town are commonly found in contaminated groundwater. An test on 39 groundwater supply in small towns conducted by the U.S. EPA, it reported that 11 VOCs could be detected in treated or untreated groundwater [12].

4.2.2. Land disposal of municipal and industrial waste

Land disposal of solid waste is the groundwater contamination source of most current concern to the general public in many developed countries and of most current regulatory interest.

Solid waste can be disposed in landfills, facilities engineered to safely contain the waste. While landfills may often prevent exposure of solid waste at the land surface, many landfills provide a direct connection with groundwater. In the past, landfill siting was based on the availability of inexpensive, undeveloped land requiring little modification for waste disposal, rather than on hydrogeologic suitability. Disposed materials often are very susceptible to leaching into groundwater.

Landfills may be grouped according to the type of materials they contain. Municipal landfills accept only non-hazardous materials, but are still likely to contain materials which pose potential health risks. Industrial landfills may contain either "hazardous" or "non-hazardous" materials. Until recently, little was known about how they were operated or what they contained. Open dumps and abandoned disposal sites generally have no engineering design. Their connection with the groundwater system and the type of materials present is often unknown. It is often in abandoned disposal sites that large volumes of highly toxic materials are found. The most hazardous solid waste disposal generally results from industrial and manufacturing activities as well as some governmental energy and defense activities. Populations of both developed and developing countries, where there is current or historical industrial activity, face potential health risks from solid waste disposal. It is reported that there will be the highest content and most types of organic contaminants in groundwater which is near the landfills. If there has 1 kilometers distance it still exist in the groundwater [13].

4.2.3. Petrochemical pollution

In recent years, there has been increasing awareness of the large number of potentially leaking underground storage gasoline tanks. For much of the twentieth century, underground storage tanks were constructed of unprotected carbon steel. Corrosion causes leaks in such tanks over some period of time, ranging from a few years to tens of years. Although the leakage from individual tanks is often small, it is often enough to contaminate a large volume of groundwater. In addition, the large number of buried tanks-several million in the United States-makes them a potentially significant groundwater contamination source. Above ground storage tanks pose less of a threat than underground tanks. Leak detection and maintenance is easier and the connection with the groundwater system is less direct. However leaks from such tanks may still act as groundwater contamination sources.

4.2.4. Agricultural activities

Numerous agricultural activities can result in non-point sources of groundwater contamination. Fertilizers, pesticides, and herbicides are applied as part of common agricultural practice throughout the world. These applications can act as sources of contamination to groundwater supplies serving large populations. Whether or not fertilizers, pesticides, and herbicides become sources of groundwater contamination depends on changing hydrogeologic conditions, application methods, and biochemical processes in the soil. In developing countries, animal and/or human waste is used for fertilizer. This is an example of the land application of sewage discussed earlier. There are the same concerns with pathogens and nitrates. The manufactured inorganic fertilizers widely used in developed countries, and finding increasing usage in all countries, also pose the threat of nitrate contamination of groundwater systems. Pesticide and herbicide application provides a source of numerous toxic organic chemicals to groundwater supplies.

Even without the introduction of fertilizers, pesticides, and herbicides, irrigation activities can lead to groundwater contamination. Naturally occurring minerals in the soil can be leached at higher rates leading to hazardous concentration levels in the groundwater. Evaporation of irrigation water can cause evaporative concentration of certain chemicals in the root zone. Flushing of these chemicals can then lead to hazardous concentration levels in groundwater.

Agricultural activities related to animals also can be groundwater contamination sources. These include the feeding of animals and the storage and disposal of their waste. Animal wastes and feedlot runoff are commonly collected in some sort of pit or tank creating the contamination threat described earlier for sewage disposal.

More than 300 pesticides were applied in Asia. The Japan is a country with the largest amount of pesticide on unit area cultivated land. Indonesia, Korea, India and China are the major consumers. But, there did not have pesticides routine monitoring in the developing countries in Asia [14].

4.2.5. Surface water and atmospheric contaminants

Groundwater is but one component of the hydrologie cycle. Groundwater quality is very much influenced by surface-water conditions and vice versa. Contamination of any surface water bodies that recharge the groundwater system is a source of groundwater contamination. This includes "natural" recharge sources such as lakes and rivers as well as "man-made" recharge sources such as artificial recharge ponds/injection wells and infiltration of urban runoff. More generally, it is important to consider the interaction of all environmental sources and pathways of pollution. Environmental contaminant sources cannot be divided into separate, isolated compartments. For example, atmospheric pollution can lead to deposition of hazardous fallout to surface waters and to soils, and eventually lead to groundwater contamination.

5. The fate and transport of chemicals in groundwater

5.1. Volatilization

Volatilization occurs in whether the vadose zone or saturated zone when the dissolved contaminants and non-aqueous phase contaminants exposed to gas. The factors affecting volatilization include solubility of the compound, molecular weight and water-saturated state of the geological media. The evaporation rate must be measured fundamentally in order to determine pollutions transporting into the atmosphere, changes of the pollution load in the vadose zone and groundwater. The process that the contaminants of deep soil volatilize to the atmosphere can be assumed as one-dimensional diffusion, which can be described with Fick's second law. Volatilization of the water-soluble organic matter, such as benzene dissolved in water is generally described by Henry's Law [15].

5.2. Adsorption

Adsorption in Soil and sediment makes an important influence on the behavior of organic pollutants. The mobility and biological toxicity reduced as organics are detained in the soil and sediment. Generally, adsorption is affected by sediments and soil properties, such as organic percentage, the type and quantity of clay minerals, cation exchange, pH and the physical and chemical properties of the contaminants. During the adsorption, the organic contaminants in the water adsorbed on the surface of the soil particles by the simultaneous distribution role of both water and solid, the driving force is mainly based on principle of "like dissolves like" and electrostatic adsorption of the polar group, and the following formulas is established [15]:

$$C_{sa} = K_a C_{wa} \tag{1}$$

The equation (1) is existed when the adsorption systems reach equilibrium. Where, Csa is the amount of organic pollutants adsorbed per unit weight of soil particles; Cwa is concentration of organic pollutants; *Ka* is the total sorption coefficient.

The adsorption of organic contaminants in soil or sediment usually described by *Ka* (soil absorption coefficient) or *Koc* (organic carbon absorption coefficient). The former refers to the ratio of the concentration of organic matter in the soil or sediment and its aqueous phase concentration. As well, the latter factor represents the ratio of the concentration of organic matter adsorbed by organic carbon in the soil or sediment and its aqueous phase concentration.

5.3. Biochemical processes

Microorganisms may play an important role in contamination transformations within groundwater and on the soil. They can act as catalysts for many types of reactions. When modeling biochemical reactions in groundwater, additional processes must be considered. These include the changes in the availability of substrate for the microorganisms to utilize and reactions on the particles that the microorganisms are attached to. When microbial reactions are significant, there is a possibility of clogging of pores due to precipitation reactions or to biomass accumulation [16].

Microorganisms not only influence chemical reactions, but may be contaminants themselves. There is much current uncertainty about the fate and survival time of viruses, bacteria, and larger enteric organisms in groundwater [17-18]. Distribution of microorganisms will vary greatly with depth. Potential outbreaks of waterborne diseases due to biologic pollutants are of particular concern where there is land disposal of human waste (often via septic tanks) and animal waste. The potential for transmittal of waterborne diseases in groundwater is particularly high in areas of rapid velocities such as karst regions.

Biodegradation mainly depends on two factors [19], the intrinsic characteristics of the pollutants (the structure of organics, physical and chemical properties) and microorganism (the activity of microbial populations), and the environmental factors controlling the reaction rate (temperature, pH, humidity, dissolved oxygen). As the U.S. Environmental Protection Agency researched [20], soil microbial degradation of organic pollutants can be expressed as a one-order response equation:

$$\frac{dc}{dt} = -KXC = -k_T c \quad (2)$$

Where, C is the mass fraction of soil organic matter[mg/g]; X is the number of active microbial in the organic matter of soil degradation[10^6 /g]; t is degradation time[d]; K is the one-order biodegradation rate constant [g/(d 10^6)]; k_r is substrate removal constant[d^{-1}].

From the above equation,

$$\frac{C}{C_0} = \exp\left(-KX_t\right) = \exp(-k_T t) \quad (3)$$

Substituted into with half-life formula,

$$t_{\frac{1}{2}} = ln2 / k_T \quad (4)$$

The half-life of degradation of residual contamination is determined.

5.4. Fate and transport in unsaturated zone [21]

In many cases, the receptor medium for release of a contaminant will be the unsaturated zone. In contrast to the saturated zone, pores in the unsaturated zone are not completely saturated with liquid. This fundamentally affects the processes governing flow and chemical transport. A number of processes will affect the contaminant within the unsaturated zone before it enters the saturated groundwater system and potentially is tapped by supply wells. The uncertainties in characterizing releases just described lead to uncertainties in defining the source terms and initial and boundary conditions for modeling unsaturated transport. Analogously, uncertainties in characterizing unsaturated transport processes lead to uncertainties in defining the source terms and initial and boundary conditions for modeling saturated transport.

For the most part, computer simulation of contaminant transport has focused on movement in the saturated zone. Assumptions are made regarding the time required for movement through the unsaturated zone. Often some sort of lag between source release and entry of chemicals into the saturated flow system is introduced into source terms. It is important to be aware of the unsaturated processes that are actually occurring, the uncertainty associated with these processes, and the role of monitoring in reducing these uncertainties.

5.5. Saturated transport [21]

Once a chemical has been released into the ground and has either moved through the unsaturated zone or directly entered the saturated zone, saturated transport processes will determine if, how fast, and at what concentration a chemical reaches a supply well. A great deal of research has been carried out on understanding and modeling these processes. There is increasing recognition that chemical transport must be viewed as a stochastic process.

The same elements of uncertainty are present for saturated transport as for unsaturated transport. The important differences are that in saturated transport, water content equals porosity, hydraulic conductivity is no longer a function of water content or head, gravity rather than suction head is the driving force, and the scale of concern may be much larger

6. Investigations for groundwater organic pollution

6.1. Current situation investigation

The current situation investigation main contents are as following:

Pollution source investigation: In groundwater polluted areas, investigate the non-point-source, line-source and point-source, and the type, pollution intensity, spatial distribution of natural source.

Investigation of unsaturated zone vulnerability: Investigate the unsaturated zone of thickness, lithological composition, composition, water permeability, the capability of degradation contaminations and so on.

Investigation of the pollution condition at the groundwater: Make sure the category, quantity or concentration of the pollutants, ascertain the pollution range, variation trend and the factors relation. All of these need samples collection in filed and laboratory test.

6.2. Pollution source investigation

With the developed, groundwater pollution attracted wide attention. In view of existing situation, we launched the survey of pollution sources, including the following aspects: industrial pollution sources, domestic pollution sources, agricultural pollution sources and surface polluted waters.

6.2.1. Industrial pollution sources

According to industrial pollution sources, we must investigate the situation as: the company name, position, sewage, waste residue (tailings) emissions, discharge, scale, pathways and outfall location, types of pollutants, quantity, composition and hazards, and the abandoned site of major polluting enterprises, abandoned wells, oil and survey of solvents and other underground storage facilities.

6.2.2. Domestic pollution sources

The survey include the distribution of dumps, scale, waste disposal methods and effects, the generation of dump leaching filtrate and components, geological structure of storage site;

amount of sewage generated, treatment and disposal of the way, main pollutants and their concentration and hazards

6.2.3. Agricultural pollution sources

Agricultural pollution sources investigation mainly include land use history and current situation, the varieties, numbers, operations, time of farmland application of chemical fertilizers and pesticides, range of sewage irrigation, main pollutants and concentration, the number of sewage irrigation and sewage irrigation amount. The scale of farms and so on.

6.2.4. Surface polluted waters

The surface polluted waters mainly about rivers, lakes, ponds, reservoirs and drains. We survey the distribution of polluted waters, the scales, the utilizations and water quality.

The coastal areas have to survey the situation of seawater invasion and saline water distribution.

7. The assessment on groundwater organic pollution

7.1. The methods of groundwater organic pollution assessment

7.1.1. The four steps of NAS

The four steps of NAS was proposed by National Academy of Sciences, United States(NAS), was an assessment method on human health risk that led by the accident, air, water, soil and other medium. The method mainly in the following aspects: the hazard identification (qualitative evaluation the degree of hazards of the chemical substances on the human health and ecological); dose-response assessment (quantitative assessment the toxicity of chemical substances, established a relationship between the dose of chemical substances and the human health hazard); exposure assessment (quantitative or qualitative estimate or calculate the exposure, exposure frequency, exposure duration and exposure mode); exposure attribute (using the data to estimate the strength of the health hazards in the different conditions or the probability of the certain health effects). This method can qualitative analysis or quantitative analysis of groundwater contamination, or combine them, the results could be quantify and analysis, and provide more detailed information to the decision-makers.

7.1.2. The four steps of EPA

In 1989, U.S. Environmental Protection Agency (EPA) promulgated the "risk assessment guidance for superfund: Human health evaluation manual", there was a similar assessment method to NAS method [22]. The steps following as data collection, exposure assessment, toxicity assessment, risk characterization. Contrast the two methods, NAS is more common methods, the use range wider, suitable for a variety of health risk assessment; the EPA method is more specific, it emphasis on the various parameters of the collection of contaminated sites, for the evaluation of contaminated sites, it more operational.

7.1.3. The MMSOILS model

The MMSOILS model is multi-media model which describe the groundwater, surface water, soil and air in the migration of chemicals, exposure and food chain accumulation [23]. Contaminate sites is multi-phase, multi-media complex. The model including the migration and transformation of pollutants module and human exposure module. Migration and transformation module include: (1)atmospheric transport pathway; (2)soil erosion; (3)groundwater migration pathway; (4)surface water pathway; (5)food chain bioaccumulation. Human exposure are: (1)adopt from drinking water, animals and plants and soil; (2)atmospheric volatiles and particulate inhalation; (3)soil, surface water and groundwater contact with skin. The model could be simulate a comprehensive migration pathway and widely used in foreign countries, and the parameters could be analysis the uncertainty.

7.1.4. The DRASTIC method

The DRASTIC method is a national standards system that developed by US EPA to evaluation aquifer vulnerability. It including: Depth to Water(D); Net Recharge(R); Aquifer media(A); Soil Media(S); Topography(T); Impact of the Unsaturated Zone Media(I); Conductivity of Aquifer Hydraulic(C). Assignment of each element from 1 to 10, and them proportional to the degree of vulnerability of groundwater. At the same time, each element is assigned a weight, the weight should be reflect the sensitivity of groundwater. The model can objectively assess the groundwater vulnerability of different areas, and its assumption that all regions of the aquifer has a uniform trend. But all the geological, hydrogeological and other conditions are different, and the model calculations defect, the DRASTIC method has some limitations.

7.2. Health–based risk assessment

7.2.1. Estimating population exposure levels

An important step in health risk assessment is the quantification of actual human exposure. Exposure can be expressed as either the total quantity of a substance that comes in contact with the human system or the rate at which a quantity of material comes in contact with the human system (mass per time or mass per time per unit body weight. The exposure assessment evaluates the type and magnitude of exposures to chemicals of potential concern at a site. The exposure assessment considers the source from which a chemical is released to the environment, the pathways by which chemicals are transported through the environmental medium, and the routes by which individuals are exposed. Parameters necessary to quantitatively evaluate dermal exposures, such as permeability coefficients, soil absorption factors, body surface area exposed, and soil adherence factors are developed in the exposure assessment. Exposure to chemicals in water can occur via direct ingestion, inhalation of vapors, or dermal absorption. Ingestion includes drinking of fluids as well as using water for rinsing and cooking of foods. Dermal absorption includes swimming and bathing.

Determination of average exposure levels for a particular population is quite difficult. This is due to difficulties in acquiring sufficient water-quality data, in identifying the exposed individuals, and in quantifying the concentrations in the different exposure pathways. For a given groundwater contamination problem, the U.S. Environmental Protection Agency stresses the importance of identifying both the currently affected population as well as possible changes in future land use. Subpopulations that may be especially sensitive to exposure should also be identified [24].

When attempting to estimate exposure to larger population entire countries, for example other concerns arise. Cothern [25] computed the average population exposure to volatile organic compounds in the United States, based on data from several thousand ground- and surface-water supplies. National exposure was estimated as a straight extrapolation of the concentration intervals from the original data. Best- and worst-case assumptions were applied for handling the "below detectable" category. Crouch *et al.* [26] applied an alternative approach to estimate population exposure levels. Rather than estimating a distribution for exposure, they made the worst-case assumption that individuals are exposed to water at the maximum measured concentration for their water supply.

7.2.2. Health risk calculations

According to the Risk Assessment Guidance for Superfund Volume I: Human Health Evaluation Manual (Part E, Supplemental Guidance for Dermal Risk Assessment) (U.S. EPA) [27], we calculation the dermal absorbed dose (DAD) and ingestion absorbed dose (IAD) [28].

$$\mathrm{DAD} = \frac{DA_{event} \times EV \times ED \times EF \times SA}{BW \times AT} \tag{5}$$

Where:

DAD=Dermally Absorbed Dose (mg/kg-day),

DA_{event}=Absorbed dose per event (mg/cm^2-event),

SA=Skin surface area available for contact (cm^2),

EV=Event frequency (events/day),

EF=Exposure frequency (days/year),

ED=Exposure duration (years),

BW=Body weight (kg),

AT=Averaging time (days).

$$IAD = \frac{\rho \times U \times EF \times ED}{BW \times AT} \tag{6}$$

Where:

IAD= Ingestion absorbed dose (mg/kg-day),

ρ= Pollutant concentration in groundwater (mg/L),

U=Drinking amount per days (L/d),

EF=Exposure frequency (days/year),

ED=Exposure duration (years),

BW=Body weight (kg),

AT=Averaging time (days).

The DAD and IAD can be represent with continuous ingestion dose (CDI).

Based on the carcinogenesis of contamination, the risk could be classified into cancer risk and noncancer hazard.

1. Noncancer hazard: Generally, the reaction of the body to non-carcinogenic substance has a dose threshold.

Lower than the threshold, they could not affect our health adversely. The non-carcinogenic risk to represent with hazard index (HI). It is defined as a ratio that continuous ingestion dose with reference dose [28].

$$\mathrm{HI} = \mathrm{CDI} / \mathrm{RfD} \tag{7}$$

Where: CDI= continuous ingestion dose (mg/kg-days), RfD= reference dose (mg/kg-days).

2. Cancer risk: There does not have dose threshold for the carcinogenic. Once it exist in environments, it will affect human health adversely. Cancer risk will be represent with risk. It is defined as a product of continuous ingestion dose with carcinogenesis slope factor.

$$\mathrm{CDI} \times \mathrm{SF} \tag{8}$$

$$1 - \exp(-\mathrm{CDI} \times \mathrm{SF}) \tag{9}$$

(If the low dose exposure risk>0.01)

Where: SF= carcinogenesis slope factor ($mg^{-1} \bullet kg \bullet d$)

When calculating the risk of a variety of substances in a variety of ways, figure out all non-cancer risk and cancer risk respectively, then add all risks together. Regardless of synergistic effect and antagonistic effect.

8. The countermeasures and suggestions

8.1. The countermeasures for groundwater pollutions

Groundwater treatment technologies are mainly as follows: pump and treat, air sparging, in-situ groundwater bioremediation and in-situ reactive walls.

8.1.1. Pump and treat technology

Pump and treat is the most common form of groundwater remediation. It is often associated with treatment technologies such as Air Stripping and Liquid-phase Granular Activated Charcoal.

Pump and treat involves pumping out contaminated groundwater with the use of a submersible or vacuum pump, and allowing the extracted groundwater to be purified by slowly proceeding through a series of vessels that contain materials designed to adsorb the contaminants from the groundwater. For petroleum-contaminated sites this material is usually activated carbon in granular form. Chemical reagents such as flocculants followed by sand filters may also be used to decrease the contamination of groundwater. Air stripping is a method that can be effective for volatile pollutants such as BTEX compounds found in gasoline.

For most biodegradable materials like BTEX, MTBE and most hydrocarbons, bioreactors can be used to clean the contaminated water to non-detectable levels. With fluidized bed bioreactors it is possible to achieve very low discharge concentrations which will meet or exceed discharge standards for most pollutants.

Depending on geology and soil type, pump and treat may be a good method to quickly reduce high concentrations of pollutants. It is more difficult to reach sufficiently low concentrations to satisfy remediation standards, due to the equilibrium of absorption (chemistry)/desorption processes in the soil.

At the figure 2, we can know how does pump and treat technology work. This system usually consists of one or more wells equipped with pumps. When the pumps are turned on, they pull the polluted groundwater into the wells and up to the surface. At the surface, the water goes into a holding tank and then on to a treatment system, where it is cleaned [29].

8.1.2. Air sparging [30]

Air sparging is an in situ groundwater remediation technology that involves the injection of a gas (usually air/oxygen) under pressure into a well installed into the saturated zone. Air sparging technology extends the applicability of soil vapor extraction to saturated soils and groundwater through physical removal of volatilized groundwater contaminants and enhanced biodegradation in the saturated and unsaturated zones. Oxygen injected below the water table volatilizes contaminants that are dissolved in groundwater, existing as a separate aqueous phase, and/or sobbed onto saturated soil particles. The volatilized contaminants migrate upward in the vadose zone, where they are removed, and generally using soil vapor extraction techniques. This process of moving dissolved and non-aqueous volatile organic compounds (VOCs), originally located below the water table, into the unsaturated zone has been likened to an in situ, saturated zone, air stripping system. In addition to this air stripping process, air sparging also promotes biodegradation by increasing oxygen concentrations in the subsurface, stimulating aerobic biodegradation in the saturated and unsaturated zones(figure 3).

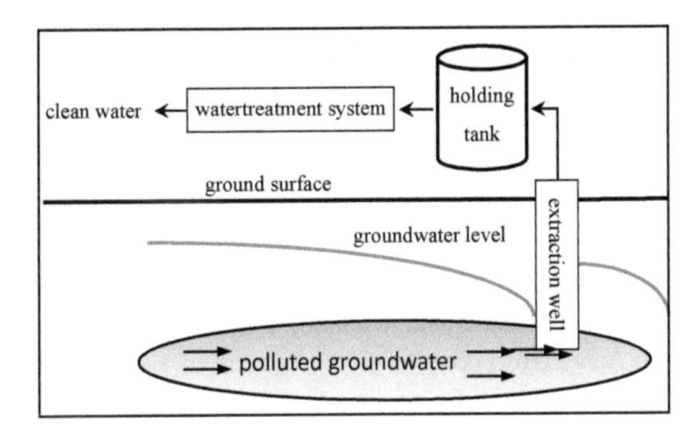

Figure 2. Pump and Treat Technology [29]

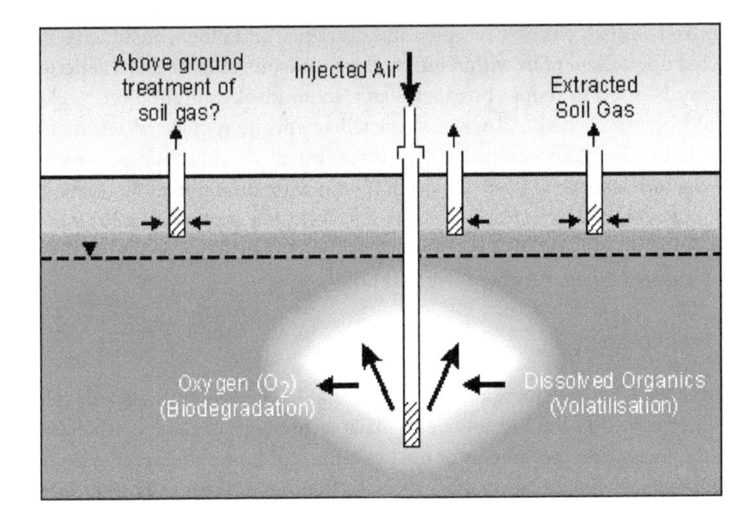

Figure 3. Air Sparging [31]

8.1.3. In-situ groundwater bioremediation [32]

In-situ groundwater bioremediation is a technology that encourages growth and reproduction of indigenous microorganisms to enhance biodegradation of organic constituents in the saturated zone. In-situ groundwater bioremediation can effectively degrade organic constituents which are dissolved in groundwater and adsorbed onto the aquifer matrix.

In-situ groundwater bioremediation can be effective for the full range of petroleum hydrocarbons. While there are some notable exceptions (e.g., MTBE) the short-chain, low-molecular-weight, more water soluble constituents are degraded more rapidly and to lower residual levels than are long-chain, high-molecular-weight, less soluble constituents. Recoverable free product should be removed from the subsurface prior to operation of the in-situ groundwater bioremediation system. This will mitigate the major source of contaminants as well as reduce the potential for smearing or spreading high concentrations of contaminants.

In-situ bioremediation of groundwater can be combined with other saturated zone remedial technologies (e.g., air sparging) and unsaturated zone remedial operations (e.g., soil vapor extraction, bioventing).

Bioremediation generally requires a mechanism for stimulating and maintaining the activity of these microorganisms. This mechanism is usually a delivery system for providing one or more of the following: An electron acceptor (oxygen, nitrate); nutrients (nitrogen, phosphorus); and an energy source (carbon). Generally, electron acceptors and nutrients are the two most critical components of any delivery system.

In a typical in-situ bioremediation system, groundwater is extracted using one or more wells and, if necessary, treated to remove residual dissolved constituents. The treated groundwater is then mixed with an electron acceptor and nutrients, and other constituents if required, and re-injected upgradient of or within the contaminant source. Infiltration galleries or injection wells may be used to re-inject treated water. In an ideal configuration, a "closed-loop" system would be established. All water extracted would be re-injected without treatment and all remediation would occur in situ. This ideal system would continually recirculate the water until cleanup levels had been achieved. If your state does not allow re-injection of extracted groundwater, it may be feasible to mix the electron acceptor and nutrients with fresh water instead. Extracted water that is not re-injected must be discharged, typically to surface water or to publicly owned treatment works (POTW).

8.1.4. In–situ reactive walls [33]

In-situ reactive walls are an emerging technology that have been evaluated, developed, and implemented only within the last few years. This technology is gaining widespread attention due to the increasing recognition of the limitations of pump and treat systems, and the ability to implement various treatment processes that have historically only been used in above-ground systems in an in situ environment. This technology is also known in the remediation industry as "funnel and gate systems" or "treatment walls".

The concept of in-situ reactive walls involves the installation of impermeable barriers downgrading of the contaminated groundwater plume and hydraulic manipulation of impacted groundwater to be directed through porous reactive gates installed within the impermeable barrier. Treatment processes designed specifically to treat the target contaminants can be implemented in these reactive or treatment gates. Treated groundwater follows its natural course after exiting the treatment gates. The flow through the treatment gates is driven by natural groundwater gradients, and hence these systems are often referred to as passive treatment walls. If a groundwater plume is relatively narrow, a permeable reactive trench can be installed across the full width of the plume, and thus preclude the necessity for installation of impermeable barriers.

In-situ reactive walls eliminate or at least minimize the need for mechanical systems, thereby reducing the long-term operation and maintenance costs that so often drive up the life cycle costs of many remediation projects. In addition, groundwater monitoring and system compliance issues can be streamlined for even greater cost savings.

Bioventing, also a modification of vapor extraction technology, is briefly contrasted with air sparging. With bioventing, extraction or injection of air into the vadose zone increases subsurface oxygen concentration, promoting bioremediation of unsaturated soil contaminants. This technique is applicable to all biodegradable contaminants, but has been applied most frequently and reportedly most successfully to sites with petroleum hydrocarbon contamination

8.2. The suggestions for groundwater pollutions

The past 40 years, groundwater subjected to pollution, it cannot be ignored that there has a serious threat to human health and ecological security problems. The research on groundwater pollution risk assessment will help understand the relationship between the soil conditions and groundwater pollution, identify the high-risk regions of groundwater pollution, provide a powerful tools for the land use and groundwater resource management, and help the policy maker and managers to develop effective management strategies and protection measures on groundwater. So we can offer some suggestions as following:

1. Continue to strengthen the research on the fate and transport in hydrogeological conditions. Hydrogeological conditions of the contaminated sites have a vital role in organic pollution of groundwater. We should pay attention to the impact that the thickness of the unsaturated zone, the aquifer lithology of unsaturated zone and groundwater, the groundwater runoff conditions on the organic pollution investigation and contaminated aquifer restoration. Unsaturated zone is the only avenue for the organic pollutant into the groundwater system. In the protection of groundwater quality, we should take impact of the physical, chemical, and biological characteristics of the unsaturated zone soil on the transport and degradation of organic pollutants into consideration.

2. In the future research, the natural attenuation of typical organic contamination in groundwater should be reinforce research, especially the organic degradation mechanism of microbes.

3. In recent years, the environmental hormone pollution research and prevention has begun to attract the attention of the world. Environmental hormone research has become the forefront and hot topic of environmental science research. But the mechanism of environmental hormone is not clearly, we should take more attention on these.

4. The research on groundwater pollution risk assessment to be carried out on the typical regions. To provide practical experience on established an reasonable and feasible groundwater pollution risk assessment system.

5. Exerting governmental function adequately and improving the laws, regulations and norms on groundwater quality monitor and assessment. Strengthening the cross-disciplinary exchanges and studies and establishing the groundwater pollution monitor network and the chemical toxicological database.

With entering a new era of environmental protection, the research of groundwater pollution risk assessment is bound to make new contributions to human survival and to protect and improve the natural environment, and to advance the theory research of environmental science.

Acknowledgements

This study is granted by the Specific Research on Public Service of Environmental Protection in China (No. 201009009). The authors appreciate the tutor and classmates for help.

Author details

Hongqi Wang*, Shuyuan Liu and Shasha Du

*Address all correspondence to: whongqi@126.com

College of Water Sciences, Beijing Normal University, Key Laboratory of Water and Sediment Sciences, Ministry of Education.Beijing, China

References

[1] Foundation for Water Research. FWR: Groundwater. http://www.euwfd.com/html/groundwater.html (accessed 18 September 2012)

[2] Barbash, J., P. V. Roberts. Volatile Organic Chemical Contamination of Groundwater Resources in the U.S. Water Pollution Control Federation 1986;58(5) 343-348.

[3] Moran, Michael J. Occurrence and Status of Volatile Organic Compounds In Ground Water From Rural, Untreated, Self-supplied Domestic Wells In the United States,

1986-99. Rapid City, SD: U.S. Department of the Interior, U.S. Geological Survey; 2002.

[4] Zoftman B C J. Persistence of organic contaminants in groundwater. In: Duivenboodne W V (ed.). Quality of Groundwater. Netherlands: Elsevier Scientific Publishing Company; 1981, p464-480.

[5] Xu Z F, Lei S H, et al. A Preliminary Study on the Analysis of Organic Pollutants in a Suburban Sewage System of Beijing. Environmental Chemistry 1988;7(1) 69-80.

[6] Liu Z C, Nie Y F, et al. The Pollution Control on Groundwater System. Beijing: China Environmental Science Press; 1991.

[7] Chen H H, He J T, Liu F, et al. Organic contamination characteristics of shallow groundwater in a study area of Taihu Lake basin, Jiangsu, China. Geological Bulletin of China 2005;24(8) 735-739

[8] Rathbun, R.E.. Transport, behavior, and fate of volatile organic compounds in streams: U.S. Geological Survey Professional Paper; 1998.

[9] Heberer T, Furhmann B, Schmidt-Baumler K, Tsipi D, Koutsouba V, Hiskia A. Occurrence of pharmaceutical residues in sewage, river, ground, and drinking water in Greece and Berlin (Germany). In: Daughton CG, Jones-Lepp TL, editors. American Chemical Society Symposium Series 791. Pharmaceuticals and personal care products in the environment. Scientific and regulatory issues; 2001. p396.

[10] Kimberlee K. B, Dana W. K, Edward T. F, et al. A national reconnaissance of pharmaceuticals and other organic wastewater contaminants in the United States -I) Groundwater. Science of the total Environment 2008;402 192-200.

[11] Liu F, Wang S M, Chen H H. Progress of investigation and evaluation on groundwater organic contaminants in western countries. Geological Bulletin of China 2010;29(6) 907-917

[12] Patrick, R.; Ford, E.; Quarles, J. Groundwater Contamination in the U.S.A.;2ed.;University of Pennsylvania Press: Philadephia; 1987. p538

[13] Zhang L.Y, Han J. L, An S.J, et al. Detection and removal of organic pollutants in refuse leaching. China Environmental Science 1998;18(2) 184-188.

[14] Hirata,T., Nakasugi, O., Yoshioka, M. et al. Groundwater pollution by volatile organochlorines in Japan and related phenomena in the subsurface environment. Water Science and Technology 1992;25(11) 9-16

[15] Li C, Wu Q. Research Advances of Groundwater Organic Contamination. Geotechnical Investigation & Surveying 2007;1 27-30.

[16] National Research Council. Groundwater Models, Scientific and Regulatory Applications. Washington, D.C., National Academy Press;1990.

[17] Keswick, B.H., Gerba, C.P. Secor, S.L., and Cech, I. Survival of enteric viruses and indicator bacteria in groundwater. Journal Environmental Science and Health 1982;17(6) 903-912.

[18] Bitton, G., Farrah, S.R., Rushkin, R.H., Butner, J. and Chou, Y.J. Survival of pathogenic and indicator organisms in groundwater. Groundwater 1983;21(4) 405-410.

[19] Zhang, W. H, Li, G. H, Shao, H.G, et al. Study on Biodegradation of Oil Pollutants on the Polluted Soils of the Aeration Zone. Research of Environmental Science 2005;15(2) 60-62.

[20] Michael, P. Bioremediation of petroleum waste from the fefining of lubricant oils. Environmental Progress 1993;12(1) 5-11.

[21] Eric R., Carl C., et al. Groundwater Contamination Risk Assessment: A Guide to Understanding and Managing Uncertainties. IAHS Press; 1990. p22-26.

[22] U.S.EPA. Risk Assessment Guidance for Superfund, Volume: Human Health Evaluation Manual (Part A). Interim final. EPA/540/l-89/002,1989.

[23] Chen, H., Liu, Z. Q, Li, G.H. A study the establishment and evaluation method of the soil risk guideline of contaminated sites. Hydrogeology & Engineering Geology 2006;21(2):84-88.

[24] U.S. EPA. Risk Assessment Guidance for Superfund, Volume 1, Human Health Evaluation Manual (Part A), Interim Final. Office of Emergency and Remedial Response, Washington, D.C.; 1989.

[25] Cothern, R.C. Techniques for the assessment of carcinogenic risk due to drinking water contaminants. CRC Critical Review of Environmental Control 1986;16(4): 357-399.

[26] Crouch, E.A.C., Wilson, R., and Zeise, L. The risks of drinking water. Water Resources Research 1983;19(6): 1359-1375.

[27] U.S.EPA. Risk Assessment Guidance for Superfund Volume I: Human Health Evaluation Manual (Part E, Supplemental Guidance for Dermal Risk Assessment) Final. EPA/540/R/99/005, 2004.

[28] Han B., He J.T, Chen H.H. Primary study of health based risk assessment of organic pollution in groundwater. Earth Science Frontiers 2006;13(1) 224-229.

[29] U.S.EPA. A Citizen's Guide to Pump and Treat. Office of Solid Waste and Emergency Response. EPA 542-F-01-025, 2001.

[30] Ralinda R. Miller, P.G. Air Sparging. Technology Overview Report, GWRTAC SERIES; 1996.

[31] Commonwealth Scientific and Industrial Research Organisation. CSIRO: Land and Water. Soil and Groundwater Remediation. http://www.clw.csiro.au/research/urban/protection/remediation/projects_2.html. (accessed 18 September 2012)

[32] U.S.EPA. Chapter X: In-Situ Groundwater Bioremediation. (ed.) How To Evaluate Alternative Cleanup Technologies For Underground Storage Tank Sites: A Guide For Corrective Action Plan Reviewers. EPA 510-R-04-002; 1995.

[33] Suthersan, S.S. IN SITU REACTIVE WALLS, Remediation engineering: design concepts, Boca Raton: CRC Press LLC; 1999.

Section 3

Treatment

Chapter 6

Advances in Electrokinetic Remediation for the Removal of Organic Contaminants in Soils

Claudio Cameselle, Susana Gouveia,
Djamal Eddine Akretche and Boualem Belhadj

Additional information is available at the end of the chapter

1. Introduction

Soil contamination is associated to industrial activities, mining exploitations and waste dumping. It is considered a serious problem since it affects not only the environment, living organisms and human health, but also the economic activities associated with the use of soil [1]. The risks associated with soil contamination and soil remediation are important points in the agenda of politicians, technicians and scientific community. The present legislation establishes a legal frame to protect the soil from potentially contaminant activities; however, the present situation of soil contamination is the result of bad practices in the past, especially related to bad waste management [2-3].

Soil contamination affects living organisms in the subsurface but also affects the plants that accumulate contaminants as they grow. Thus, contaminants enter the food chain with a potential impact in public health [4]. On the other hand, contaminants can be washed out the soil by rain and groundwater, resulting in the dissemination of the contamination. This process is not desirable because the area affected by the contaminants is bigger and bigger and the possible remediation is more difficult and costly as the affected area grows [5]. Therefore, soil contamination is a serious problem that requires a rapid solution in order to prevent more environmental damages. Prevention is the best "technology" to save our soils from the contamination. A strict management of the wastes and good environmental practices associated to industrial activities, mining, transportation and dumping management are required to prevent the contamination of the environment. However, many sites have been identified as contaminated sites. The European Union, USA, Canada, Japan and South Korea made a lot of efforts in recent years to identify the contaminated sites in each country. The new legislation, especially in the European Union, forces the administration to identify the contaminated sites and evaluate the risks associated to the environment and public health. Then, the remediation of those sites must be carried out, starting with the riskier sites for humans and living organisms [6]. This is the aim of the present legislation in Spain about the management of wastes and soil contamination [7, 8] which is the transposition of the European Directive 2008/98/CE [9].

Soil remediation implies the application of a technology able to remove or eliminate the contaminants following by the restoration of the site to the original state. So far, it sounds easy to do. However, there is not a technology able to remove any kind of contaminants in any kind of soil. Moreover, the restoration of the site to the original state is not always possible due to the characteristics of the soil and/or the remediation technology. Thus, the common objective in soil remediation is to remove the contaminants to a safe level for humans and the environment, and restore the properties of the soil to a state appropriate for the common soil uses [10-12]. So, the final target concentration to consider the soil non-contaminated will be different depending on the future use of the soil: urban, agriculture or industrial.

During the last 20 years, scientist and technicians spent a lot of efforts in the developing of innovative technologies for soil remediation [13]. Those technologies use the physical, chemical and biological principles to remove and/or eliminate the contaminants from soil. Thus, for instance, bioremediation uses the capacity of soil microorganisms to degrade organic contaminants into the soil [14]. Thermal desorption was designed to remove volatile and semi-volatile organics. Gasoline, BETX, chlorinated organics can be removed by thermal desorption, but also PAHs or PCBs [15]. Soil washing uses a solution in water to dissolve the contaminants from soil. Once the soil is clean, it can be stored in the same place and the contaminants will undergo a stabilization process [16]. Soil remediation technologies can be applied in situ, i.e. in the contaminated site without excavation, or ex-situ: the soil is excavated and it is treated in a facility specifically designed for the remediation process. In situ technologies are preferred because they results in lower costs, less exposition to the contaminants and less disruption of the environment. However, the control of the operation is more difficult and depending on the permeability of the soil and soil stratification, the operation may results in very poor results. On the other hand, ex-situ technologies permit a better control of the operation, and the remediation results are not very affected by some soil characteristics as permeability and stratification [17-19].

2. Electrokinetic remediation: Basis and applications

Electrokinetic remediation is an environmental technique especially developed for the removal of contaminants in soil, sediments and sludge, although it can be applied to any solid porous material [20]. Electrokinetic remediation is based in the application of a direct electric current of low intensity to the porous matrix to be decontaminated [21]. The effect of the electric field induces the mobilization and transportation of contaminants through the porous matrix towards the electrodes, where they are collected, pumped out and treated. Main electrodes, anode and cathode, are inserted into the soil matrix, normally inside a chamber which is fill with water or the appropriate solution to enhance the removal of contaminants (Figure 1). Typically, a voltage drop of 1 VDC/cm is applied to the main electrodes.

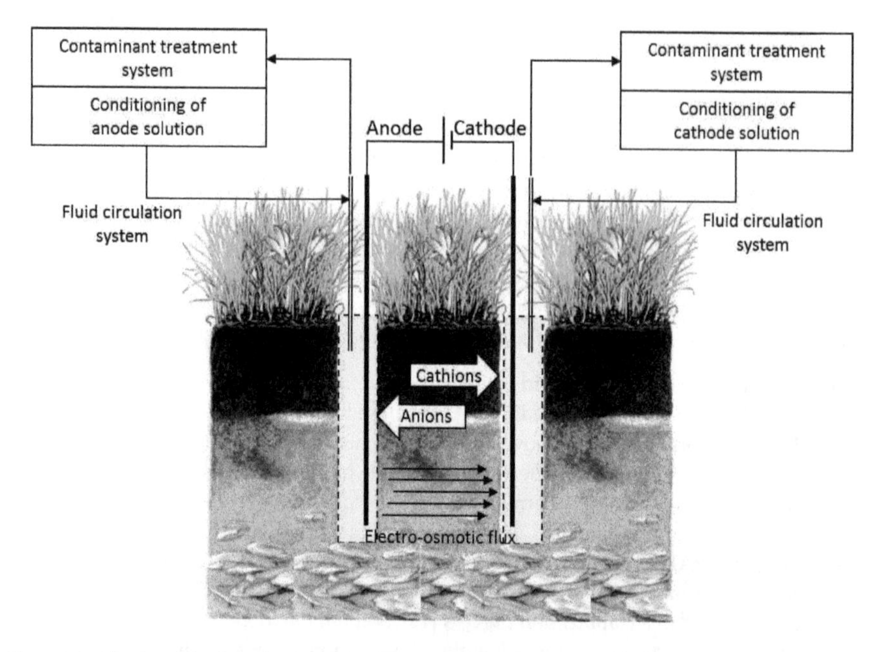

Figure 1. Application of the electrokinetic remediation in a contaminated site.

Contaminants are transported out of the soil due several transportation mechanisms induced by the electric field [22, 23]:

- Electromigration is defined as the transportation of ions in solution in the interstitial fluid in the soil matrix towards the electrode of the opposite charge (Figure 2). Cations move toward the cathode (negative electrode), and anions move toward the anode (positive electrode). The ionic migration or electromigration depends on the size and charge of the ion and the strength of the electric field.

- Electro-osmosis is the net flux of water or interstitial fluid induced by the electric field (Figure 2). Electro-osmosis is a complex transport mechanism that depends on the electric characteristics of the solid surface, the properties of the interstitial fluid and the interaction between the solid surface and the components in solution. The electro-osmotic flow transports out of the porous matrix any chemical species in solution. Soils and sediments are usually electronegative (solid particles are negatively charged), so the electro-osmotic flow moves toward the cathode. In the case of electropositive solid matrixes, the electro-osmotic flow moves toward the anode. Detailed information about electro-osmosis can be found in literature [24].

- Electrophoresis is the transport of charged particles of colloidal size and bound contaminants due to the application of a low direct current or voltage gradient relative to the stationary pore fluid. Compared to ionic migration and electro-osmosis, mass transport by electrophoresis is negligible in low permeability soil systems. However, mass transport by electrophoresis may become significant in soil suspension systems and it is the mechanism for the transportation of colloids (including bacteria) and micelles.

- Diffusion refers to the mass transport due to a concentration gradient, not to a voltage gradient as the previous transport mechanisms. During the electrokinetic treatment of contaminated soils, diffusion will appear as a result of the concentration gradients generated by the electromigration and electro-osmosis of contaminants. Diffusive transport is often neglected considering its lower velocity compared to electromigration and electro-osmosis.

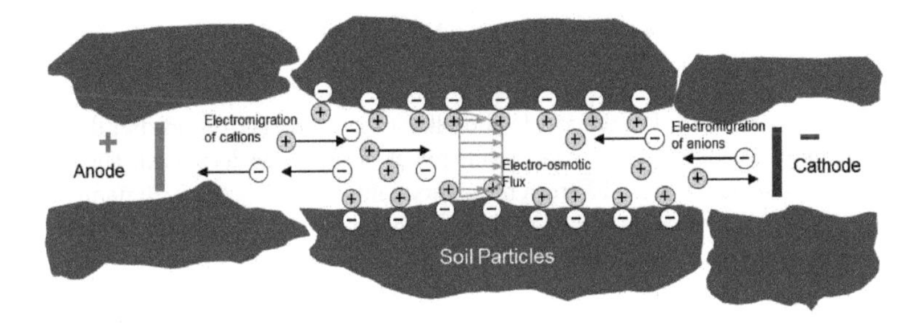

Figure 2. Transport mechanisms in electrokinetic remediation

The two main transport mechanisms in electrokinetic remediation are electromigration and electro-osmosis [25]. The extent of electromigration of a given ion depends on the conductivity of the soil, soil porosity, pH gradient, applied electric potential, initial concentration of the specific ion and the presence of competitive ions. Electromigration is the major transport processes for ionic metals, polar organic molecules, ionic micelles and colloidal electrolytes.

The electro-osmotic flow depends on the dielectric constant and viscosity of pore fluid as well as the surface charge of the solid matrix represented by zeta potential. The zeta potential is a function of many parameters including the types of clay minerals and ionic species that are present as well as the pH, ionic strength, and temperature. Electro-osmosis is considered the dominant transport process for both organic and inorganic contaminants that are in dissolved, suspended, emulsified or such similar forms. Besides, electro-osmotic flow though low permeability regions is significantly greater than the flow achieved by an ordinary hydraulic gradient, so the electro-osmotic flow is much more efficient in low permeability soils [26].

The application of an electric field to a moisten porous matrix also induces chemical reactions into the soil and upon the electrodes that decisively influences the chemical transportation and speciation of the contaminants and other constituents of the soil. Chemical reactions include acid-alkaline reactions, redox reaction, adsorption-desorption and dissolution-precipitation reactions. Such reactions dramatically affect the speciation of the contaminants and therefore affect the transportation and contaminant removal efficiency [27]. The main reaction in the electrochemical/electrokinetic systems is the decomposition of water that occurs at the electrodes. The electrolytic decomposition of water reactions generates oxygen gas and hydrogen ions (H^+) due to oxidation at the anode and hydrogen gas and hydroxyl (OH^-) ions due to reduction at the cathode as shown in equations 1 and 2.

At Anode (Oxidation):

$$2\,H_2O \rightarrow 4\,e^- + 4H^+_{(aq)} + O_{2(gas)} \quad E^0 = -1.229\ V \tag{1}$$

At Cathode (Reduction):

$$4\,H_2O + 4\,e^- \rightarrow 2\,H_{2(gas)} + 4\,OH^-_{(aq)} \quad E^0 = -0.828\ V \tag{2}$$

Essentially, acid is produced at the anode and alkaline solution is produced at the cathode, therefore, pH in the cathode is increased, while pH at the anode is decreased. The migration of H^+ ions from the anode and OH^- from the cathode into the soil leads to dynamic changes in soil pH. H^+ is about twice as mobile as OH^-, so the protons dominate the system and an acid front moves across the soil until it meets the hydroxyl front in a zone near the cathode where the ions may recombine to generate water. Thus, the soil is divided in tow zones with a sharp pH jump in between: a high pH zone close to the cathode, and a low pH zone on the anode side. The actual soil pH values will depend on the extent of transport of H^+ and OH^- ions and the geochemical characteristics of the soil. The implications of these electrolysis reactions are enormous in the electrokinetic treatment since they impact the absorption/desorption of the contaminants, the dissolution/precipitation reactions, chemical speciation and the degradation of the contaminants. Moreover, pH changes into the soil affects the contaminant migration, and the evolution of the electro-osmotic flow which is decisive in the removal of non-charged organic contaminants [20]. In electrokinetic remediation, it is also common the use of chemical to enhance the dissolution and the transportation of the contaminants. The enhancing chemical are going to interact with the soil and the contaminants, therefore it is necessary to evaluate the geochemistry of the soil and the possible reactions with the enhancing chemicals, considering at the same time the effect of the pH, in order to design a satisfactory technique that removes or eliminates the contaminants keeping the natural properties of the soil for its use after the remediation process.

3. Removal of organic contaminants by electrokinetics: Limitations and enhancements

Electrokinetic remediation was first proposed and tested for the removal of heavy metals and other charged inorganic contaminants in soils, sediments and sludges. However, the electrokinetic remediation is also useful for the removal or elimination of organic contaminants [28]. Considering the different physico-chemical properties of the organic contaminants compared to the properties of heavy metals, the operating conditions of the electrokinetic treatment and the enhancing chemicals will be rather different than those used for heavy metal polluted soils. The main transportation mechanisms in electrokinetic remediation are: electromigration and electro-osmosis. In general, the more dangerous and persistent organic contaminants are not soluble in water (which is the interstitial fluid in natural soils) and are neither ionic nor ionizable molecules. Therefore, electromigration cannot be considered as the transport mechanisms for organic contaminants. Electro-osmosis is the net flux of water in the soil matrix that flows through the soil from one electrode to the other due to the effect of the electric field. Electro-osmotic flow moves towards the cathode in electronegatively charged soils, which is the most common case. Again, organic contaminants are not soluble in water and therefore their elimination from soils cannot be achieved in an unenhanced electrokinetic treatment. In order to achieve an effective removal or elimination of organic contaminants from soils, their solubility in has to be enhanced with the use of co-solvents, surfactants or any other chemical agent. Alternatively, the removal or elimination of organic contaminants can be achieved by the combination of electrokinetics and other remediation techniques such as chemical oxidation/ reduction, permeable reactive barriers, electrolytic reactive barriers or thermal treatment. For the removal of organic contaminants, both solubilization of the contaminants and adequate electro-osmotic flow are required, which appear to be quite challenging to accomplish simultaneously. The electro-osmotic flow is found to be dependent on the magnitude and mode of electric potential application. The electro-osmotic flow is higher initially under higher electric potential, but it reduces rapidly in a short period of time. Interestingly, the use of effective solubilizing agent (surfactant) and periodic voltage application was found to achieve the dual objectives of generating high and sustained electro-osmotic flow and at the same time induce adequate mass transfer into aqueous phase and subsequent removal. Periodic voltage application consists of a cycle of continuous voltage application followed by a period of "down time" where the voltage was not applied was found to allow time for the mass transfer, or the diffusion of the contaminant from the soil matrix, to occur and also polarize the soil particles. Several laboratory studies have demonstrated such desirable electro-osmotic flow behavior in a consistent manner, but field demonstration projects are needed to validate these results under scale-up field conditions [29, 30].

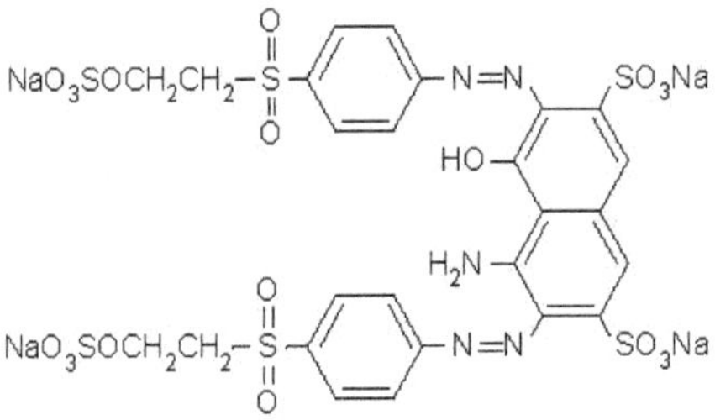

Figure 3. Chemical structure of reactive black 5

3.1. Electrokinetic removal of soluble organics

Although most dangerous organic contaminants in soils, sediments and sludges are persistent hydrophobic organics, several works in literature focused on the treatment of soils with soluble organics. Thus, reactive black 5 is a common dye used in the industry. Reactive black 5 is a complex organic molecule difficult to biodegrade in the environment and shows a significant toxicity for living organisms in soils and water. Reactive black 5 is soluble in water, but it can be retained in soils and sediments adsorbed upon the surface of mineral particles and organic matter. Considering the chemical structure of the reactive black 5 (figure 3), the molecule can be ionized at alkaline pH when the sulfonic groups are neutralized forming an anion with 4 negative charges. In this conditions, reactive black 5 can be transported by electrokinetics toward the anode, but only if the molecule is in solution. The desorption of the molecule can be achieved using potassium sulfate as flushing solution in the anode and cathode chambers. Figure 4 shows the advance of the Reactive Black 5 toward the anode by electromigration. The advance of the day is evident in the 4th day of treatment, and it is completely removed from the soil in 5 days. The removal of Reactive black 5 is only possible when the molecule is desorbed from the soil particles but the electromigration was only possible when the pH into the soil was alkaline [31]. The pH was controlled in the anode (the left hand side in figure 4) at a value below 7 and the alkaline front electrogenerated at the cathode (the right hand side in figure 4) advanced through the soil favoring the dissolution and electromigration of reactive black 5. Negligible Reactive Black 5 was observed if the pH into the soil was not alkaline.

3.2. Co-solvents

Most of organic contaminants of environmental concern are practically insoluble in water but they can be dissolved in other organic solvents. Thus, the use of other processing fluid than water may help in the desorption and dissolution of the organic contaminants in soils, sediments and sludges. Electrokinetic remediation is an in situ technique, and water is always present in soils. So, the organic solvent will not be used alone but in combination with water as a co-solvent. Thus, the possible organic solvents to be used are now reduced to those miscible with water. But this is not the unique condition a co-solvent has to meet. Organic co-solvents have to be safe for the environment or with a minor environmental impact, and it has to be easy to recover from soil after the treatment. Apart from the environmental limitations in the selection of the co-solvents, there are also some technical aspects to take into account. The use of co-solvents mixed with water decreases the conductivity of the processing fluid due to the decrease of salts solubility in the organic co-solvent. It decreases the current intensity through the soil. The presence of an organic co-solvent will also affects the viscosity of the processing fluid and change the interaction between the processing fluid and the soil particles. Those alterations will impact directly the evolution of the electro-osmotic flow which is the main transportation mechanism for the removal of organic contaminants. Any rate, the increase in the contaminant solubility due to the use of the co-solvent may largely compensate the decrease in the electroosmotic flow, being the result positive for the removal of the organic contaminants. Some of the co-solvents used in literature are: ethanol, n-butanol, n-butylamine, tetrahydrofuran, or acetone [26, 32-34]. Phenanthrene was the target contaminant in the studies with co-solvents. The removal of phenanthene was negligible when water was used as flushing solution but the removed fraction of phenanthrene clearly increased with the use of co-solvents, especially n-butylamine which resulted in a removal of 43% in 127 days in a lab test with a soil specimen of 20 cm long. The removal can be enhanced controlling other variables such as the pH into the soil and improving the electro-osmotic flow operating at higher voltage gradient or with periodic voltage application [34].

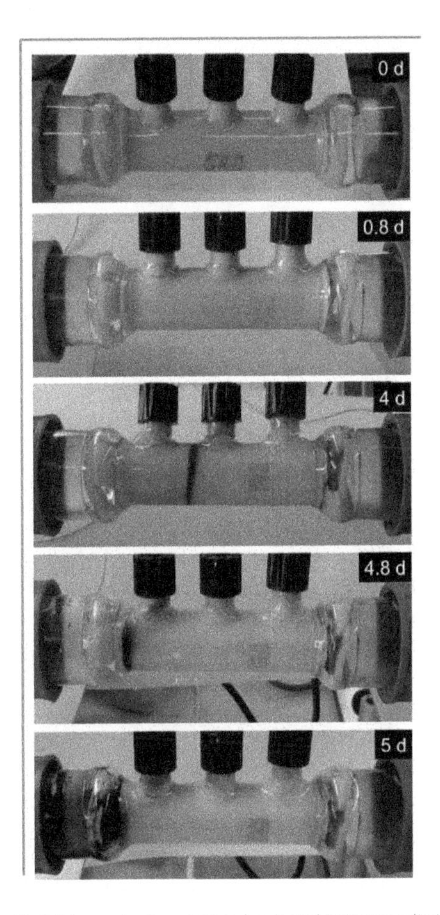

Figure 4. Removal of Reactive Black 5 from a kaolin specimen by electrokinetic remediation

3.3. Surfactants

The name surfactant is the short version of "surface-active agent". It means that the so-called surfactants are a group of substances that has in common a special capacity to change the surface properties of the solution when they are present. In environmental applications, the interest of surfactants is their ability to lower the surface and interfacial tension of water improving the solubility of hydrophobic organics. There is a wide variety of chemical structures and families that fits in the definition of surfactant. Basically a surfactant is a chemical compound whose molecule includes a hydrophilic group in one side of the molecule and in the opposite side a hydrophobic group or chain. The interaction of the hydrophilic group with water assures its solubility whereas the interaction of the hydrophobic group with the organic contaminants assures the solubilization of hydrophobic organics. The hydrophobic group or chain in the surfactant molecule is repelled by water, so the surfactant molecules tend to form spherical structures with the hydrophilic group outside and the hydrophobic chains inside. These spherical structures are called micelles. Thus, the surfactant creates a hydrophobic environment very appropriate for the solubilization of organic compounds. The formation of micelles depends on the surfactant concentration and the micelle formation reach a maximum for a surfactant concentration called CMC "critical micelle concentration" [26].

There is a wide variety of chemical structures in the surfactants, but usually they are classified by the electric charge in the molecule in 4 groups: cationic, anionic, neutral and zwitterionic (includes positive and negative charges in the same molecule). In environmental applications, neutral or anionic surfactants are preferred because cationic surfactants tend to interact with the soil particles, retarding their advance and reducing their effectiveness [26]. The toxicity of surfactants to the soil microorganisms it is also very important for the remediation and restoration of soils. That is why in recent years the research was redirected to the use of natural surfactants or biosurfactants [35].

A wide variety of surfactants have been used in electrokinetic remediation for the removal of organic contaminants: Sodium dodecyl sulfate (SDS), Brij 35, Tween 80, Igepal CA-720, Tergitol and other. Target contaminant in these studies includes hydrophobic and persistent organics such as: phenanthrene, DDT, diesel, dinitrotoluene, hexachlorobenzene and others. In general, it can be conclude that the reported results in literature are quite good reaching removal efficiencies over 80% in many studies, at least in bench scale laboratory test with both model and real contaminated soils [36, 37]. Reddy et al. demonstrated the removal of phenanthrene by electrokinetics using surfactants as an enhanced flushing solution in the electrode chambers. Different types of soils, commonly kaolin and glacial till, were used in this study. In general, there is no elimination of phenanthrene when water was used as flushing solution despite the large electro-osmotic flow registered in these experiments. The use of surfactants such as Igepal CA-720, Tween 80 or Witconol tend to decrease the electro-osmotic flow due to the changes in the interaction of the flushing solution with the soil particle surface, the decreasing in the electric conductivity of the system, and the increase of the viscosity of the flushing solution. Despite the decreasing of the electro-osmotic flow, the increase of phenanthrene solubility in the surfactant flushing solution resulted in a very important transportation and removal of phenanthrene in the fluid collected on the cathode side. The specific removal results did not only depend on the type and concentration of surfactant but also in the pH evolution into the soil, the type of soil and the ionic strength in the processing fluid. Those variables affect the solubilization of the organic contaminants by the surfactant, but the main influence is in the develonment and evolution of the electro-osmotic flow. Thus, the limitation of very acidic environments into the soil avoids a sharp reduction of the electro-osmotic flow. This can be achieved controlling the pH on the anode or using a buffering solution with the flushing surfactant solution. The buffering capacity of the soil also contributes to avoid the acidification of the interstitial fluid [26, 33, 38, 39]. However, the only use of surfactants seems to be not enough to get a complete removal of phenanthrene from polluted soils, and it is necessary to enhance the electro-osmotic flow using high voltage gradients (2 V/cm or higher) and even the use of periodic voltage applications operating with a constant voltage drop intermittently. The periodic voltage application resulted in about 90% of the phenanthrene removed on the cathode solution [40].

3.4. Cyclodextrins

Glucose may form cyclic structures with 6, 7 or 8 molecules called cyclodextrins. The resulting molecule has the structure of a truncated cone. The internal cavity has different size depending on the number of glucose units. The inner diameter of the molecule ranged from 0.45-0.53 nm for α-cyclodextrin (ring of 6 glucose molecules); 0.60-0.65 nm for β-cyclodextrin (ring of 7 glucose molecules); and 0.75-0.85 nm for γ-cyclodextrin (ring of 8 glucose molecules). Cyclodextrin shows an amphiphilic behavior due to the rings of –OH groups present at the both ends of the molecule. The hydroxyl groups are polar and confer to the cyclodextrin the solubility in water. However, the inner surface of the molecule is hydrophobic and cyclodextrins can accommodate different non-polar, hydrophobic molecules such as aliphatic, aromatic or lipophilic compounds. Moreover, the different size of the inner cavity of the ciclodextrin molecules can be used as a select the molecules to be trapped inside, and therefore, transported and removed.

Cyclodextrins have been used to enhance the removal of hydrophobic organics such as phenanthrene [41], dinitrotoluene [42], the herbicide atrazine [43], and other contaminants [44] in real and model soils. In general, cyclodextrins are facilitating agents that improve the removal of organic contaminants from soil compared to other experiments with unenhanced electrokinetics, but results from cyclodextrin tests are usually less effective than test with surfactants, iron nanoparticles or with chemical oxidants. The efficiency of the removal can be enhanced combining more than one facilitating agent. Thus, Pham et al. [45] and Oonnittan et al. [46] used the electrokinetic treatment with a cyclodrextring flushing solution, combined with ultrasounds or chemical oxidation with hydrogen peroxide. Anyway, the use of cyclodextrins may enhance the removal of the hydrophobic contaminants but the results are usually lower than that found with surfactants.

4. Combined technologies

4.1. Electrokinetics and chemical oxidation/reduction

Electrokinetic remediation is a technique that removes the contaminants from the contaminated soil by transportation (electro-osmosis and electromigration). However, organic contaminants are difficult to remove from soils, mainly due to the low solubility in water, and their strong adsorption to organic matter and soil particles. There are some other ways to look at the problem of organic contaminants in soil. One possibility is to degrade the contaminants in situ. To achieve such degradation, it is necessary to create the adequate conditions into the soil supplying strong oxidizing chemicals to the soil pores to perform the degradation in situ. Oxidants such as ozone, hydrogen peroxide or persulfate can be transported into de soil by electromigration and/or electro-osmosis. As the oxidants advance through the soil, they react with the organic contaminants resulting in smaller molecules usually less toxic that the original ones. The objective is to be able to completely oxidize the organic contaminants to carbon dioxide and water. If such complete degradation is not possible under the operating conditions into the soil, the formation of simpler molecules are considered enough, because small and simpler molecules can be degraded easily by the microorganisms into the soil. Thus, this technology can be a very attractive solution for the degradation of complex organic contaminants into soil. This technology does not generate waste effluents with harmful compounds because they are destroyed into the soil. Moreover, the contact of the workers with the contaminants and contaminated soil particles are reduced to a minimum, which is a very important point in the field operation.

On the other hand, the chemical destruction of organic contaminants can be carried out by chemical reduction, when a reductive chemical process results in less toxic compounds. Thus, organochlorine pesticides can be degraded by reductive dechlorination. The result is the organic molecule without chlorine atoms in its structure. Thus, the resulted organic compounds are much less toxic than the original compound and they can be easily degraded by the microorganisms into the soil.

There are several applications of chemical oxidation combined with electrokinetics in literature. Yukselen-Aksoy and Reddy [47] have tested the degradation of PCB in contaminated soils by persulfate. Sodium persulfate is a strong oxidizing agent with a standard reduction potential of 2.7 V which assures the effective oxidation of most of the organic contaminants. Persulfate is firstly transported into the soil by electromigration and/or electro-osmosis, and then it is activated by pH or temperature. To active the persulfate, it is necessary to achieve over 45ºC or acidify the soil below 4. Both conditions can be reached with the electric field. High voltage gradient results in the heating of soil; and the acid front electrogenerated at the anode can acidify the soil. So, in this case the application of the electric field not only was used as a transportation mechanism but as a tool to control the key variables of the process. In this work [47], the highest degradation of PCBs was achieved in a kaolin specimen with a 77.9% of removal when temperature was used as activator of the persulfate.

The combination of electrokinetics and chemical oxidation was tested in a contaminated soil with hexachlorobenzene [46, 48]. Hydrogen peroxide was supplied to the soil from the anode in a Fenton-like process where the iron content in the soil was sufficient to activate the descomposition of H_2O_2 for the generation of hydroxyl radicals (OH). 60% of HCB was eliminated from the soil in 10 days of treatment avoiding the deactivation of the Fenton reagent at high pH values. Higher removal can be achieved at longer treatment time, controlling the pH in the optimum range for Fenton reagent which is slightly acid environments. At alkaline pH, H_2O_2 decomposes in water and oxygen and do not form OH radicals.

The use of Fe^0 for the remediation of soils has been used recently for the ability of the native iron to catalyze the reductive dechlorination of organic compounds such as pentachlorophenol, trichloroethylene, hexachlorobenzene and others. In this technology, the electric field can be used as a driving force to transport the nanoparticles into the soil. Reddy and Karri [49] found that the combination of electrokinetic remediation and Fe0 nanoparticles can be applied for the removal of pentachlorophenol from soil. The transportation of Fe^0 nanoparticles was determined by the iron concentration into the soil at the end of the experiments. Iron concentration at the end of the experiments increased with the initial Fe^0 concentration used in the anode and with the voltage gradient. However, the transport of nanoparticles was limited by their aggregation, settlement and partial oxidation within the anode. Pentachlorophenol was partially reduced (40-50%) into the soil, but a complete PCP elimination was found near the cathode due to the combination of Fe^0 and the reductive dechlorination within the cathode. In order to favor the transportation of nanoparticles into the soil, new strategies are needed to prevent aggregation, settlement and oxidation of iron nanoparticles for enhanced remediation of soils. Cameselle et al. [50] studied the surface characteristics of the iron nanoparticles and proposed several dispersant to favor the transportation and avoid aggregation and settlement. Among the dispersants proposed aluminum lactate presents good characteristics to be used in large scale application. Other metallic catalysts such as Cu/Fe or Pd/Fe bimetal microscale particles were satisfactorily used for the remediation of soils with organochlorines.. Dechlorination of hexachlorobenzene up to 98% was achieved with Cu/Fe [51] and only 60% with Pd/Fe [52].

4.2. Electrokinetics and permeable reactive barriers

Permeable reactive barriers (PRB)are passive remediation systems especially designed for the remediation of contaminated ground water. PRBs consist of digging a trench in the path of flowing groundwater and then filling it with a selected permeable reactive material. As the contaminated groundwater passes through the PRB, contaminants react with the active material in the PRB being absorbed, precipitated or degraded. Clean groundwater exits the PRB. In the design of a PRB several factors have to be taking into account. First, the nature and the chemical properties of the contaminants have to be considered for the selection of the reactive material. For organic contaminants, materials such as active carbon or Fe^0 were used. Organic contaminants can be retained in in the porous structure of the active carbon. Native iron has been used for the reductive dechlorination of pesticides and other organochlorines. The flow rate of groundwater and the reaction rate of the contaminants with the active material in the PRB are used to define the width of the barrier. The resident time of the groundwater in the barrier has to be enough to reach a complete removal or degradation of the contaminants. Finally, the porous structure of the barrier has to confer the barrier itself a permeability value higher than the surrounding soil, to assure that all the groundwater pass through the barrier and there will not be bypass. The main advantages of the PRB are the stable operation for long treatment time, even several years, with very low investment and maintenance costs. Anyway, the limited results found in several application impulse the research in several directions in order to improve the removal of the contaminants [53]. One possibility is the combination of the PRB with electrokinetic remediation.

The combination of electrokinetic remediation with PRB has been satisfactory used to remediate soils polluted with heavy metals such as chromium. The electric field transports the chromium towards the main electrodes, but in their way, the chromium ions pass through a PRB made of elemental iron. The chemical reduction of chromium takes places reacting with the elemental iron. The electric field also plays a role in the reduction of the chromium [54]. In the case of organic contaminants, Chang and Cheng [55] applied the combination of PRB with electrokinetics to remediate a soil specimen contaminated with perchloroethilene. The experiments were carried out at a constant voltage drop of 1 v/cm and sodium carbonate 0.01 M was used as processing fluid to avoid the formation of an acid front in the anode. It eliminates the acidification of the soil and the possible negative effects on the electro-osmotic flow. The PRB were made of nanoparticles of elemental iron and zinc. The perchloroethylene is dechlorinated upon the nanoparticles of iron and zinc. However, the formation of ferric oxide and ferric hydroxides limits the activity of the PRB and its operational life. The protons electrogenerated at the cathode can contribute in the solubilization and removal of the ferric hydroxides increasing the activity and duration of the PRB. Moreover, the proton also favors the dechlorination reaction. In conclusion, the formation of H+ ions in the anode favors the elimination of perchloroethylene. As the voltage drop applied to the system increases, the formation of H+ upon the anode also increases resulting in more and faster perchloroethylene removal. Thus, the operation at 2 V/cm resulted in the removal of almos 99% of the initial perchloroethylene in only 10 days of operation.

Chung and Lee [56] applied a combination of electrokinetics with PB for the remediation of the tetrachloroethylene contaminated soils and groundwater. The interest of this work is the media used in the PRB. The authors used a mixture of sand with a material they called atomizing slag (material patented) which is basically a mixture of oxides of Si, Fe, Ca and Al. The atomizing slag is mainly used as a construction material but it was selected for the PRB because is much cheaper than other materials reported in literature such as iron nanoparticles. The operation of such system in situ resulted in the removal of 90% of the tetrachloroethylene considering the concentrations measured before and after the system electrokinetic-PRB confirmeing the suitability of this technology for its application in situ to contaminated soils.

4.3. Bioelectroremediation

The combination of electrokinetic remediation with bioremediation has shown some interesting results that promise this technology a good development in the near future. Basically, the application of an electric field to a polluted site may help in the mobilization of the contaminants. That mobilization makes the contaminants available for the microorganisms. At the same time, soil bacteria are like a colloid with a surface charge. So, they can be moved under the effect of the electric field. The transport of bacteria, even in small distances, may help in the interaction between the bacteria and the contaminants. Finally, the electric field can be used as a transportation mechanism to introduce into the soil the nutrients and other chemicals that may facilitate the bacterial growth and development, as well as the supply of other chemicals that can contribute to the degradation of the contaminants [57, 58].

Lageman [59] developed a technology called Electrokinetic Biofence (EBF). The aim of the EBF is to enhance biodegradation of the VOCs in the groundwater at the zone of the fence by electrokinetic dispersion of the dissolved nutrients in the groundwater. EBF which consists of a row of alternating cathodes and anodes with a mutual distance of 5 m. Upstream of the line of electrodes, a series of infiltration wells were installed, which have been periodically filled with nutrients. After running the EBF for nearly 2 years, clear results have been observed. The concentration of nutrients in the zone has increased, the chloride index is decreasing, and VOCs are being dechlorinated by bio-activity. The electrical energy for the EBF is being supplied by solar panels.

4.4. Electroheating

The removal of volatile and semi-volatile organics from soil can be carried out heating the soil, evaporating the volatile organics and aspirating the vapors, which in turn are trapped in the appropriate absorbent such as active carbon to be finally eliminated by incineration. The heating of soil can be done in several ways. One possibility is the use of an electric current. Soil is not a good electric conductor, so the passing of an electric current generates heat. In electrokientic remediation, it is used a continuous electric current because the objective is to transport the ionic and nonionic contaminants out of the soil. In the case of electroheating, a transportation of the contaminants using the electric field as a driven force is not necessary. The electric field is only used as a source of energy that is transformed from electric energy into heat. That is why in electroheating the continuous electric field is substituted by an alternate current that supplies the energy but does not induce transportation. Soil is not a good electric conductor. The conductivity of soils is much lower than the typical electric conductor such as metals. The conductivity of soil largely varies with the moisture content and the presence of mobile ions. Anyway, the conductivity of soil is usually low and the heating is easy to achieve with an alternate current. It is recommendable to avoid the use of electroheating in saturated soils. A soil saturated in moisture favors the transportation of current instead of the electric heating.

Electroheating shows several advantages form other technologies designed for the removal of volatile organics from soils. In electroheating, the heating of soil is directly related to the electric field intensity. So, the increase of temperature and the final temperature in the soil can be easily controlled adjusting the intensity of the electric field. Furthermore, the heat is generated into the soil, in the whole volume at the same time, achieving a more uniform temperature in the area to be treated. The uniform temperature permits a uniform removal of the contaminants and a more efficient use of the energy.

Electrical heating was used in the remediation of a contaminated site in Zeist, the Netherlands [60]. The site was severely polluted with chlorinated solvents such as perchloroethylene (PCE) and trichloroethylene (TCE) and their degradation products are cis-1,2-dichloroethene (C-DCE) and vinyl chloride (VC). Satisfactory results were obtained in the application of electrical heating soil and groundwater in the source areas, combined with soil vapor extraction and low-yield groundwater pumping, and enhancing biodegradation in the groundwater plume area. Two years of heating and 2.5 years of biodegradation has been resulted in near-complete removal of the contaminants. A full scale implementation of six phase electrical heating technology was used in Sheffield, UK [61]. Terra Vac Ltd. demonstrated how remediation timescales can be reduced from months/years to weeks, with an electrical heating capable of remediation of soil in difficult geological conditions and in dense populated urban areas. TCE and VC were remediated by electrical heating up to 99.99%. Smith [62] applied the electroheating technology for the remediation of dicholormethane, ethylene dibromide, triclhoroethane and tetrachloroethane. Electroheating was an effective technology for the remediation of such organic contaminants, but during the remediation process, the elevation of temperature increases the solubility of the contaminants in the groundwater, the activity of soil microorganisms is enhanced and some reactions, such as hydrolysis of the contaminants and the desorption of gases, takes place. Those factors may affect the removal of the contaminants and it influence has to be considered.

5. Large scale applications

Electrokinetic remediation has been used as a remediation technology in several tests at field scale. In the USA, field projects were carried out or funded by USEPA, DOE, ITRC, US-Army Environmental Centre, as well as companies like Electropetroleum Inc. [63], Terran Corporation, and Monsanto, Dupont, and General Electric which developed the Lasagna™ technology [64, 65]. In Europe, more field projects with electrokinetic remediation have been carried out, specially associated to the commercial activity of the Hak Milieutechniek Company [66, 67]. Recently, some field experiences were reported in Japan and Korea [68]. Some of these tests deal about the remediation of polluted sites with organic contaminants such as organochlorides, PAHs and PCBs. Considering the information available in literature, the cost of field application of electrokinetic remediation is about an average value of 200 $\$/m^3$ for both organic and inorganic contaminants, however it must be kept in mind that electrokinetic remediation, like any other remediation technology, is site specific and the costs can be vary from less than 100 to more than 400 $\$/m^3$ [69].

6. Future perspectives

The scientific knowledge accumulated in the last 20 years conducted to several lessons learned that must be keep in mind in the design of projects for the remediation of contaminated sites. Thus, the remediation of contaminated soils with organic contaminants is site specific. The results obtained in the remediation of a site cannot be assumed for other contaminated sites. This is due to the large influence of the physicochemical properties of the soil and its possible interactions with the organic contaminants in the results of the electrokinetic remediation treatment. Besides, the chemicals used for enhancing the electrokinetic treatment may complicate the behavior of the system and the removal results may largely vary from one site to another. Recently, it has been considered that the combination of several remediation techniques may improve the remediation results, especially in sites with complex contamination, including recalcitrant organics compounds and inorganic contaminants. The combination of electrokinetics with bioremediation, phytoremediation, chemical oxidation or electrical heating, presents very interesting perspectives for the remediation of difficult sites. It is expected the combination of remediation technologies to improve the remediation results, saving energy and time.

Author details

Claudio Cameselle[1], Susana Gouveia[1*], Djamal Eddine Akretche[2] and Boualem Belhadj[2*]

*Address all correspondence to: gouveia@uvigo.es

*Address all correspondence to: belhadj_b@hotmail.fr

1 Department of Chemical Engineering, University of Vigo, Building Fundicion, Vigo, Spain

2 Laboratory of Hydrometallurgy and Inorganic Molecular Chemistry, Faculty of Chemistry, USTHB, BP 32, El- Alia, Bab Ezzouar, Algiers, Algeria

References

[1] Wania F. On the origin of elevated levels of persistent chemicals in the environment. Environmental Science and Pollution Research 1999;6(1):11-19.

[2] Basheer C, Obbard JP, Lee HK. Persistent organic pollutants in Singapore's coastal marine environment: Part II, sediments. Water Air Soil Pollut 2003;149(1-4):315-325.

[3] Galassi S, Provini A, De Paolis A. Organic micropollutants in lakes: A sedimentological approach. Ecotoxicol Environ Saf 1990;19(2):150-159.

[4] Ramakrishnan B, Megharaj M, Venkateswarlu K, Sethunathan N, Naidu R. Mixtures of environmental pollutants: Effects on microorganisms and their activities in soils. Reviews of Environmental Contamination and Toxicology 2011;211:63-120.

[5] Vaudelet P, Schmutz M, Pessel M, Franceschi M, Guérin R, Atteia O, et al. Mapping of contaminant plumes with geoelectrical methods. A case study in urban context. J Appl Geophys 2011;75(4):738-751.

[6] Rodrigues SM, Pereira ME, da Silva EF, Hursthouse AS, Duarte AC. A review of regulatory decisions for environmental protection: Part I - Challenges in the implementation of national soil policies. Environ Int 2009;35(1):202-213.

[7] REAL DECRETO 9/2005, de 14 de enero, por elque se establece la relación de actividadespotencialmente contaminantes del suelo y loscriterios y estándares para la declaración desuelos contaminados. BOE 2005;15:1833-1843. Accessible in: http://www.boe.es/boe/dias/2005/01/18/pdfs/A01833-01843.pdf

[8] Ley 22/2011, de 28 de julio, de residuos y suelos contaminados.BOE 2011;181: 85650-85705. Accessible in: http://www.boe.es/boe/dias/2011/07/29/pdfs/BOE-A-2011-13046.pdf

[9] Directive 2008/98/EC of the european parliament and of the councilof 19 November 2008on waste and repealing certain Directives. Official Journal of the European Union L312:3-30. Accessible in: http://eur-lex.europa.eu/LexUriServ/LexUriServ.do?uri=OJ:L:2008:312:0003:0030:en:PDF

[10] Karn B, Kuiken T, Otto M. Nanotechnology and in situ remediation: A review of the benefits and potential risks. Ciencia e Saude Coletiva 2011;16(1):165-178.

[11] Rayu S, Karpouzas DG, Singh BK. Emerging technologies in bioremediation: constraints and opportunities. Biodegradation 2012:1-10.

[12] Tong M, Yuan S. Physiochemical technologies for HCB remediation and disposal: A review. J Hazard Mater 2012;229-230:1-14.

[13] USEPA. Remediation technologies screening matrix and reference guide. 2012. Accessible in: http://www.frtr.gov/matrix2/section1/toc.html

[14] Garbisu C, Alkorta I. Bioremediation: Principles and future. Journal of Clean Technology, Environmental Toxicology and Occupational Medicine 1997;6(4):351-366.

[15] Baker RS, LaChance J, Heron G. In-pile thermal desorption of PAHs, PCBs and dioxins/furans in soil and sediment. Land Contamination and Reclamation 2006;14(2): 620-624.

[16] Sierra C, Gallego JR, Afif E, Menéndez-Aguado JM, González-Coto F. Analysis of soil washing effectiveness to remediate a brownfield polluted with pyrite ashes. J Hazard Mater 2010;180(1-3):602-608.

[17] Romantschuk M, Sarand I, Petänen T, Peltola R, Jonsson-Vihanne M, Koivula T, et al. Means to improve the effect of in situ bioremediation of contaminated soil: An overview of novel approaches. Environmental Pollution 2000;107(2):179-185.

[18] Suer P, Nilsson-Paledal S, Norrman J. LCA for site remediation: A literature review. Soil and Sediment Contamination 2004;13(4):415-425.

[19] Mulligan CN, Yong RN, Gibbs BF. An evaluation of technologies for the heavy metal remediation of dredged sediments. J Hazard Mater 2001;85(1-2):145-163.

[20] Reddy KR, Cameselle C. Overview of electrochemical remediation technologies. In: Reddy KR, Cameselle C. (Eds). Electrochemical remediation technologies for polluted soils, sediments and groundwater. Wiley. New Jersey. USA. 2009. pp:3-28.

[21] Acar YB. Principles of electrokinetic remediation. Environment Science Technology 1993;27(13):2638-2647.

[22] Acar YB, Gale RJ, Putnam GA, Hamed J, Wong RL. Electrochemical processing of soils: Theory of pH gradient development by diffusion, migration, and linear convection. Journal of Environmental Science and Health - Part A Environmental Science and Engineering 1990;25(6):687-714.

[23] Acar YB, Gale RJ, Alshawabkeh AN, Marks RE, Puppala S, Bricka M, et al. Electrokinetic remediation: Basics and technology status. J Hazard Mater 1995;40(2):117-137.

[24] Probstein RF, Hicks RE. Removal of contaminants from soils by electric fields. Science 1993;260(5107):498-503.

[25] Pamukcu S. Electrochemical transport and transportation In: Reddy KR, Cameselle C. (Eds). Electrochemical remediation technologies for polluted soils, sediments and groundwater. Wiley. New Jersey. USA. 2009. pp:29-64.

[26] Saichek RE, Reddy KR. Electrokinetically enhanced remediation of hydrophobic organic compounds in soils: A review. Crit Rev Environ Sci Technol 2005;35(2):115-192.

[27] Yeung A.T. (2009). Geochemical processes affecting electrochemical remediation In: Reddy KR, Cameselle C. (Eds). Electrochemical remediation technologies for polluted soils, sediments and groundwater. Wiley. New Jersey. USA. 2009. pp:65-94.

[28] Pamukcu S. Electro-chemical technologies for in-situ restoration of contaminated subsurface soils. Electronic Journal of Geotechnical Engineering 1996;1.

[29] Reddy KR, Darko-Kagya K, Al-Hamdan AZ. Electrokinetic remediation of pentachlorophenol contaminated clay soil. Water Air Soil Pollut 2011;221(1-4):35-44.

[30] Cameselle C, Reddy KR. Development and enhancement of electro-osmotic flow for the removal of contaminants from soils. Electrochim Acta 2012.

[31] Pazos M, Ricart MT, Sanromán MA, Cameselle C. Enhanced electrokinetic remediation of polluted kaolinite with an azo dye. Electrochim Acta 2007;52(10 SPEC. ISS.): 3393-3398.

[32] Li A, Cheung KA, Reddy KR. Cosolvent-enhanced electrokinetic remediation of soils contaminated with phenanthrene. J Environ Eng 2000;126(6):527-533.

[33] Reddy KR, Saichek RE. Effect of soil type on electrokinetic removal of phenanthrene using surfactants and cosolvents. J Environ Eng 2003;129(4):336-346.

[34] Saichek RE, Reddy KR. Effect of pH control at the anode for the electrokinetic removal of phenanthrene from kaolin soil. Chemosphere 2003;51(4):273-287.

[35] Seyed Razavi SN, Khodadadi A, Ganjidoust H. Treatment of soil contaminated with crude-oil using biosurfactants. Journal of Environmental Studies 2012;37(60):107-116.

[36] Gomes HI, Dias-Ferreira C, Ribeiro AB. Electrokinetic remediation of organochlorines in soil: Enhancement techniques and integration with other remediation technologies. Chemosphere 2012;87(10):1077-1090.

[37] Yeung AT, Gu Y-. A review on techniques to enhance electrochemical remediation of contaminated soils. J Hazard Mater 2011;195:11-29.

[38] Saichek RE, Reddy KR. Surfactant-enhanced electrokinetic remediation of polycyelic aromatic hydrocarbons in heterogeneous subsurface environments. Journal of Environmental Engineering and Science 2005;4(5):327-339.

[39] Saichek RE, Reddy KR. Evaluation of surfactants/cosolvents for desorption/solubilization of phenanthrene in clayey soils. Int J Environ Stud 2004;61(5):587-604.

[40] Reddy KR, Saichek RE. Enhanced Electrokinetic Removal of Phenanthrene from Clay Soil by Periodic Electric Potential Application. Journal of Environmental Science and Health - Part A Toxic/Hazardous Substances and Environmental Engineering 2004;39(5):1189-1212.

[41] Maturi K, Reddy KR. Simultaneous removal of organic compounds and heavy metals from soils by electrokinetic remediation with a modified cyclodextrin. Chemosphere 2006;63(6):1022-1031.

[42] Khodadoust AP, Reddy KR, Narla O. Cyclodextrin-enhanced electrokinetic remediation of soils contaminated with 2,4-dinitrotoluene. J Environ Eng 2006;132(9): 1043-1050.

[43] Wang G, Xu W, Wang X, Huang L. Glycine-β-cyclodextrin-enhanced electrokinetic removal of atrazine from contaminated soils. Environ Eng Sci 2012;29(6):406-411.

[44] Reddy KR, Ala PR, Sharma S, Kumar SN. Enhanced electrokinetic remediation of contaminated manufactured gas plant soil. Eng Geol 2006;85(1-2):132-146.

[45] Pham TD, Shrestha RA, Sillanpää M. Removal of hexachlorobenzene and phenanthrene from clayey soil by surfactant- and ultrasound-assisted electrokinetics. J Environ Eng 2010;136(7):739-742.

[46] Oonnittan A, Shrestha RA, Sillanpää M. Effect of cyclodextrin on the remediation of hexachlorobenzene in soil by electrokinetic Fenton process. Separation and Purification Technology 2009;64(3):314-320.

[47] Yukselen-Aksoy Y, Reddy KR. Effect of soil composition on electrokinetically enhanced persulfate oxidation of polychlorobiphenyls. Electrochim Acta 2012.

[48] Oonnittan A, Shrestha RA, Sillanpää M. Remediation of hexachlorobenzene in soil by enhanced electrokinetic Fenton process. Journal of Environmental Science and Health - Part A Toxic/Hazardous Substances and Environmental Engineering 2008;43(8):894-900.

[49] Reddy KR, Karri MR. Effect of electric potential on nanoiron particles delivery for pentachlorophenol remediation in low permeability soil. Proceedings of the 17th In-

ternational Conference on Soil Mechanics and Geotechnical Engineering: The Academia and Practice of Geotechnical Engineering; 2009.

[50] Cameselle C, Reddy K, Darko-Kagya K, Khodadoust A.Effect of Dispersant on Transport of Nanoscale Iron Particles in Soils: Zeta Potential Measurements and Column Experiments. J Environ Eng 2013 (in press).

[51] Zheng Z, Yuan S, Liu Y, Lu X, Wan J, Wu X, et al. Reductive dechlorination of hexachlorobenzene by Cu/Fe bimetal in the presence of nonionic surfactant. J Hazard Mater 2009;170(2-3):895-901.

[52] Wan J, Li Z, Lu X, Yuan S. Remediation of a hexachlorobenzene-contaminated soil by surfactant-enhanced electrokinetics coupled with microscale Pd/Fe PRB. J Hazard Mater 2010;184(1-3):184-190.

[53] Henderson AD, Demond AH. Long-term performance of zero-valent iron permeable reactive barriers: A critical review. Environ Eng Sci 2007;24(4):401-423.

[54] Weng C-, Lin Y-, Lin TY, Kao CM. Enhancement of electrokinetic remediation of hyper-Cr(VI) contaminated clay by zero-valent iron. J Hazard Mater 2007;149(2): 292-302.

[55] Chang JH, Cheng SF. The remediation performance of a specific electrokinetics integrated with zero-valent metals for perchloroethylene contaminated soils. J Hazard Mater 2006;131(1-3):153-162.

[56] Chung HI, Lee M. A new method for remedial treatment of contaminated clayey soils by electrokinetics coupled with permeable reactive barriers. Electrochim Acta 2007;52(10 SPEC. ISS.):3427-3431.

[57] Lohner ST; Tiehm A, Jackman SA, CarterP. (2009). Coupled electrokinetic-bioremediation: applied aspects. In: Reddy KR, Cameselle C. (Eds). Electrochemical remediation technologies for polluted soils, sediments and groundwater. Wiley. New Jersey. USA. 2009. pp: 389-416.

[58] Wick, L.Y. (2009). Coupling electrokinetics to the bioremediation of organic contaminants: principles and fundamental interactions. In: Reddy KR, Cameselle C. (Eds). Electrochemical remediation technologies for polluted soils, sediments and groundwater. Wiley. New Jersey. USA. 2009. pp:369-387.

[59] Lageman, R., Pool, W (2009). Electrokinetic biofences. In: Reddy KR, Cameselle C. (Eds). Electrochemical remediation technologies for polluted soils, sediments and groundwater. Wiley. New Jersey. USA. 2009. pp:357-366.

[60] Lageman R, Godschalk MS. Electro-bioreclamation. A combination of in situ remediation techniques proves successful at a site in Zeist, the Netherlands. Electrochim Acta 2007;52(10 SPEC. ISS.):3449-3453.

[61] Fraser, A. (2009). Remediation of contaminated site using electric resistive heating. First use in the UK. Symposium on electrokinetic remediation, EREM 2009. Lisbon (Portugal). pp: 47-48.

[62] Smith, GJ. (2009). Coupled electrokinetic-thermal desorption. In: Reddy KR, Cameselle C. (Eds). Electrochemical remediation technologies for polluted soils, sediments and groundwater. Wiley. New Jersey. USA. 2009. pp: 505-536.

[63] Wittle JK., Pamukcu, S., Bowman, D., Zanko, L.M., Doering, F. (2009). Field studies of sediment remediation. In: Reddy KR, Cameselle C. (Eds). Electrochemical remediation technologies for polluted soils, sediments and groundwater. Wiley. New Jersey. USA. 2009. pp: 661-696.

[64] Ho SV, Athmer C, Sheridan PW, Hughes BM, Orth R, McKenzie D, et al. The lasagna technology for in situ soil remediation. 1. Small field test. Environmental Science and Technology 1999;33(7):1086-1091.

[65] Ho SV, Athmer C, Sheridan PW, Hughes BM, Orth R, Mckenzie D, et al. The lasagna technology for in situ soil remediation. 2. Large field test. Environmental Science and Technology 1999;33(7):1092-1099.

[66] Lageman, R. Pool, W. (2009). Experiences with field applications of electrokinetic remediation. In: Reddy KR, Cameselle C. (Eds). Electrochemical remediation technologies for polluted soils, sediments and groundwater. Wiley. New Jersey. USA. 2009. pp: 697-718.

[67] Godschalk MS, Lageman R. Electrokinetic Biofence, remediation of VOCs with solar energy and bacteria. Eng Geol 2005;77(3-4 SPEC. ISS.):225-231.

[68] Chung, H.I., Lee, M.H.(2009). Coupled electrokientic PRB for remediation of metals in groundwater. In: Reddy KR, Cameselle C. (Eds). Electrochemical remediation technologies for polluted soils, sediments and groundwater. Wiley. New Jersey. USA. 2009. pp: 647-660.

[69] Athmer, CJ.(2009) Costs estimates for electrokinetic remediaiton. In: Reddy KR, Cameselle C. (Eds). Electrochemical remediation technologies for polluted soils, sediments and groundwater. Wiley. New Jersey. USA. 2009. pp: 583-588.

Chapter 7

Adsorption Technique for the Removal of Organic Pollutants from Water and Wastewater

Mohamed Nageeb Rashed

Additional information is available at the end of the chapter

1. Introduction

Organic pollution is the term used when large quantities of organic compounds. It originates from domestic sewage, urban run-off, industrial effluents and agriculture wastewater. sewage treatment plants and industry including food processing, pulp and paper making, agriculture and aquaculture. During the decomposition process of organic pollutants the dissolved oxygen in the receiving water may be consumed at a greater rate than it can be replenished, causing oxygen depletion and having severe consequences for the stream biota. Wastewater with organic pollutants contains large quantities of suspended solids which reduce the light available to photosynthetic organisms and, on settling out, alter the characteristics of the river bed, rendering it an unsuitable habitat for many invertebrates. Organic pollutants include pesticides, fertilizers, hydrocarbons, phenols, plasticizers, biphenyls, detergents, oils, greases, pharmaceuticals, proteins and carbohydrates [1-3].

Toxic organic pollutants cause several environmental problems to our environment. The most common organic pollutants named persistent organic pollutants (POPs). POPs are compounds of great concern due to their toxicity,persistence, long-range transport ability [4] and bioaccumulation in animals [5], travel long distances and persist in living organisms. POPs are carbon-based chemical compounds and mixtures (twelve pollutants) that include industrial chemicals such as polychlorinated biphenyls (PCBs),polychlorinated dibenzo-p-dioxins and dibenzofurans (PCDD/Fs), and some organochlorine pesticides (OCPs), such as hexachlorobenzene (HCB) or dichloro-diphenyl-trichloroethane (DDT), dibenzo-p-dioxins (dioxins) and dibenzo-p-furans (furans) [6]. PCDD/Fs are released to the environment as by-products of several processes, like waste incineration or metal production [7]. Many of these compounds have been or continue to be used in large quantities and due to their environmental persistence, have the ability to bioaccumulate and biomagnify [8].

Efficient techniques for the removal of highly toxic organic compounds from water have drawn significant interest. A number of methods such as coagulation, filtration with coagulation, precipitation, ozonation, adsorption, ion exchange, reverse osmosis and advanced oxidation processes have been used for the removal of organic pollutants from polluted water and wastewater. These methods have been found to be limited, since they often involve high capital and operational costs. On the other hand ion exchange and reverse osmosis are more attractive processes because the pollutant values can be recovered along with their removal from the effluents. Reverse osmosis, ion exchange and advanced oxidation processes do not seem to be economically feasible because of their relatively high investment and operational cost.

Among the possible techniques for water treatments, the adsorption process by solid adsorbents shows potential as one of the most efficient methods for the treatment and removal of organic contaminants in wastewater treatment. Adsorption has advantages over the other methods because of simple design and can involve low investment in term of both initial cost and land required. The adsorption process is widely used for treatment of industrial wastewater from organic and inorganic pollutants and meet the great attention from the researchers. In recent years, the search for low-cost adsorbents that have pollutant –binding capacities has intensified. Materials locally available such as natural materials, agricultural wastes and industrial wastes can be utilized as low-cost adsorbents. Activated carbon produced from these materials can be used as adsorbent for water and wastewater treatment [9].

2. Adsorption phenomenon

Adsorption is a surface phenomenon with common mechanism for organic and inorganic pollutants removal. When a solution containing absorbable solute comes into contact with a solid with a highly porous surface structure, liquid–solid intermolecular forces of attraction cause some of the solute molecules from the solution to be concentrated or deposited at the solid surface. The solute retained (on the solid surface) in adsorption processes is called adsorbate, whereas, the solid on which it is retained is called as an adsorbent. This surface accumulation of adsorbate on adsorbent is called adsorption. This creation of an adsorbed phase having a composition different from that of the bulk fluid phase forms the basis of separation by adsorption technology.

In a bulk material, all the bonding requirements (be they ionic, covalent, or metallic) of the constituent atoms of the material are filled by other atoms in the material. However, atoms on the surface of the adsorbent are not wholly surrounded by other adsorbent atoms and therefore can attract adsorbates. The exact nature of the bonding depends on the details of the species involved, but the adsorption process is generally classified as physicsorption (characteristic of weak Van Der Waals forces) or chemisorption (characteristic of covalent bonding). It may also occur due to electrostatic attraction.

As the adsorption progress, an equilibrium of adsorption of the solute between the solution and adsorbent is attained (where the adsorption of solute is from the bulk onto the adsorbent is minimum). The adsorption amount (qe, mmol g^{-1}) of the molecules at the equilibrium step was determined according to the following equation:

$$qe = V(Co\text{-}Ce)\ /M \tag{1}$$

where V is the solution volume (L); M is the mass of monolithic adsorbents (g); and Co and Ce are the initial and equilibrium adsorbate concentrations, respectively.

Other definition of adsorption is a mass transfer process by which a substance is transferred from the liquid phase to the surface of a solid, and becomes bound by physical and/or chemical interactions. Large surface area leads to high adsorption capacity and surface reactivity [10].

2.1. Adsorption isotherms and models

An adsorption isotherm is the presentation of the amount of solute adsorbed per unit weight of adsorbent as a function of the equilibrium concentration in the bulk solution at constant temperature. Langmuir and Freundlich adsorption isotherms are commonly used for the description of adsorption data.

The Langmuir equation is expressed as:

$$Ce\,/\,qe = \ 1\,/\,bXm + Ce\,/\,Xm, \tag{2}$$

Where Ce is the equilibrium concentration of solute (mmol L^{-1}), qe is the amount of solute adsorbed per unit weight of adsorbent (mmol g^{-1} of clay), Xm is the adsorption capacity (mmol g^{-1}), or monolayer capacity, and b is a constant (L $mmol^{-1}$).

The Freundlich isotherm describes heterogeneous surface adsorption. The energy distribution for adsorptive sites (in Freundlich isotherm) follows an exponential type function which is close to the real situation. The rate of adsorption/desorption varies with the strength of the energy at the adsorptive sites. The Freundlich equation is expressed as:

$$logqe = \ logk + \ 1\,/\,n\ logCe\ , \tag{3}$$

Where k (mmol g^{-1}) and $1/n$ are the constant characteristics of the system [11].

3. Types of adsorbents

Different types of adsorbents are classified into natural adsorbents and synthetic adsorbents. Natural adsorbents include charcoal, clays, clay minerals, zeolites, and ores. These natural materials, in many instances are relatively cheap, abundant in supply and have significant potential for modification and ultimately enhancement of their adsorption capabilities. Synthetic adsorbents are adsorbents prepared from Agricultural products and wastes, house hold wastes, Industrial wastes, sewage sludge and polymeric adsorbents. Each adsorbent has its own characteristics such as porosity, pore structure and nature of its adsorbing surfaces. Many waste materials used include fruit wastes, coconut shell, scrap tyres, bark and other tannin-rich materials, sawdust, rice husk, petroleum wastes, fertilizer wastes, fly ash, sugar industry wastes blast furnace slag, chitosan and seafood processing wastes, seaweed and algae, peat moss, clays, red mud, zeolites, sediment and soil, ore minerals etc.

Activated carbons as adsorbent for organic pollutants consists in their adsorption a complex process and there still exists considerable difficulty. The main cause of this difficulty results from the large number of variables involved. These include, for example, electrostatic, dispersive and chemical interactions, intrinsic properties of the solute (for example solubility and ionization constant), intrinsic properties of the adsorbent (such pore size distribution), solution properties (in particular, pH) and the temperature of the system [12].

Activated carbons (AC) (both granular activated carbon (GAC) and powdered activated carbons (PAC)) are common adsorbents used for the removal of undesirable odor, color, taste, and other organic and inorganic impurities from domestic and industrial waste water owing to their large surface area, micro porous structure nonpolar character and due to its economic viability.The major constituent of activated carbon is the carbon that accounts up to 95% of the mass weight In addition, active carbons contain other hetero atoms such as hydrogen, nitrogen, sulfur, and oxygen. These are derived from the source raw material or become associated with the carbon during activation and other preparation procedures [13-14]. Putra et al. [15] investigated the removal of Amoxicillin (antibiotic) from pharmaceutical effluents using bentonite and activated carbon as adsorbents. The study was carried out at several pH values. Langmuir and Freundlich models were then employed to correlate the equilibria data on which both models fitted the data equally well. While chemisorption is the dominant adsorption mechanism on the bentonite, both physicosorption and chemisorption played an important role for adsorption onto activated carbon.

Adsorption of methane on granular activated carbon (GAC) was studied. The results showed that with decreasing temperature or increasing methane uptake by GAC the adsorption efficacy decreased. Interactions between the methane molecules and the surface of carbon increase the density of adsorbed methane in respect to the density of compressed gas. The effect that the porosity and the surface chemistry of the activated carbons have on the adsorption of two VOC (benzene and toluene) at low concentration (200 ppm) was also studied. The results show that the volume of narrow micropores (size <0.7 nm) seems to govern the adsorption of VOC at low concentration, specially for benzene adsorption. AC with low content in oxygen surface groups has the best adsorption capacities. Among the AC tested, those prepared by chemical activation with hydroxides exhibit the higher adsorption capacities for VOC. The adsorption capacities achieved are higher than those previously shown in the literature for these conditions, especially for toluene. Adsorption capacities as high as 34 g benzene/100 g AC or 64 g toluene/100 g AC have been achieved [16].

4. Adsorption of dyes

Adsorption techniques are used as high quality treatment processes for the removal of dissolved organic pollutants, such as dyes, from industrial wastewater. Dyes consider as type of organic pollutants. The textile, pulp and paper industries are reported to utilize large quantities of a number of dyes, these pollutant may be found in wastewaters of many industries generating considerable amounts of colored wastewaters, toxic and even carcinogenic, posing serious hazard to aquatic living organisms. Dyes represent one of the problematic groups; they are emitted into wastewater from various industrial branches, mainly from the dye manufacturing and textile finishing and also from food coloring, cosmetics, paper and carpet industries. It is well known that the dye effluents from dyestuff manufacturing and textile industries, may exhibit toxic effects on microbial populations and can be toxic and/or carcinogenic to mammalian animal. Most dyes used in textile industries are stable to light and are not biologically degradable. Furthermore, they are resistant to aerobic digestion. [17].

On searching for economical and available starting materials; different low cost adsorbents were used for the removal of dyes. Activated rice husk was used as cheap adsorbent for color removal from wastewater [18]. Hamdaoui [19] reported that the maximum adsorption of basic dye, methylene blue, onto cedar sawdust and crushed brick was 60 and 40 mg L^{-1}, respectively. Wood-shaving bottom ash (WBA) was used for the removal of Red Reactive 141 (RR141), and azo reactive dyes. WBA/H_2O and WBA/H_2SO_4 adsorbents were made by treating WBA with water and 0.1 M H_2SO_4, respectively; to increase adsorption capacity. The effects of different parameters on adsorption (effect of contact time, initial pH of solution, dissolved metals and elution) were studied.The maximum dye adsorption capacities of WBA/H_2O and WBA/H_2SO_4 obtained from a Langmuir model at 30 °C were 24.3, 29.9, and 41.5 mg l^{-1}, respectively. In addition, WBA/H_2O and WBA/H_2SO_4 could reduce colour and high chemical oxygen demand (COD) of real textile wastewater [20]. Beer brewery waste has been shown to be a low-cost adsorbent for the removal of methylene blue dye from the aqueous solution. The results of preliminary adsorption kinetics showed that the diatomite waste could be directly used as a potential adsorbent for removal of methylene blue on the basis of its adsorption–biosorption mechanisms [21].

Sewage sludge was applied for the preparation of activated carbon adsorbent. Activated carbon adsorbent prepared from sewage sludge has being identified as a potentially attractive material for wastewater. Research studies has been conducted to demonstrate the uses of treated sewage sludge for the removal of dyes from wastewater and polluted water [22-27]. Otero et al. [27] produced activated carbon by chemically activation and pyrolysis of sewage sludge. The properties of this type of material was studied by liquid-phase adsorption using crystal violet, indigo carmine and phenol as adsorbates. Three prepared activated carbon,of different particle sizes, were used ASS-g1 (particle diameter<0.12 mm), ASS-g2 (0.12<particle diameter<0.5 mm) and PSS-g2(0.12<particle diameter<0.5 mm). Crystal violet dye adsorption has been higher (Qmax 263.2 mg/g using AAS, 270 mg/g using ASS and 184 mg/g using PPS) than indigo carmine (Qmax 60.04 mg/g using AAS, 54.8 mg/g using ASS and 30.8 mg/g using PPS). They proposed that activated carbons made from sewage sludge show promise for the removal of organic pollutants from aqueous streams.

Dye	Adsorbent	Adsorption capacity	Reference
Reactive Blue 2	activated carbon	0.27,mmol/g	[31]
Reactive Red 4,	activated carbon	0.24 mmol/g	[31]
Reactive Yellow 2	activated carbon	0.11 mmol/g	[31]
Everzol Black B	Sepiolite	120.5 g/kg	[32]
Everzol Red 3BS	Sepiolite	108.8 g/kg	[33]
Everzol Red 3BS	Zeolite	111.1 g/kg	[32]
Everzol Black B	Zeolite	60.6 g/kg	[32]
Orange-G	bagasse fly ash	1.245 g/kg	[33]
Methyl Violet	bagasse fly ash	3.712 g/kg	[33]
Acid Blue 113	amino-functionalized nanoporous silica SBA-3	769 g/kg	[34]
Acid Red 114	amino-functionalized nanoporous silica SBA-3	1000 g/kg	[34]
Acid Green 28	amino-functionalized nanoporous silica SBA-3	333 g/kg	[34]
Acid Yellow 127	amino-functionalized nanoporous silica SBA-3	1250 g/kg	[34]
Acid Orange 67	amino-functionalized nanoporous silica SBA-3	2500 g/kg	[34]
Acid Blue 25	waste tea activated carbon	203.34 mg/g	[35]
Methylene Blue	bituminous coal-based activated carbon	580 mg/g	[36]
Methylene Blue	coal-based activated carbon coal-based activated carbon(KOH washed)	(252 mg/g 234.0 mg/g	[37]
Methylene Blue	activated carbon from Cotton stalk-based	180.0 mg/g	[38]
Methylene Blue	activated carbon from Posidonia oceanica (L.) dead leaves:	285.7 mg/g	[39]
Methylene Blue	Salix psammophila activated carbon	225.89 mg/g	[40]

Dye	Adsorbent	Adsorption capacity	Reference
Methylene Blue	activated carbon from flamboyant pods (Delonix regia)	890mg/g	[41]
Methylene Blue	activated carbon from from Oil palm wood-based	90.9 mg/g	[42]
Methylene Blue	activated carbon from Oil palm shell-based	243.9 mg/g	[43]

Table 1. Selected adsorbents used for dyes removal from polluted water

The utilization of fly ash in removing dyes from textile wastewater was investigated. A calcium-rich fly ash had been used as adsorbent for the removal of Congo red dye under different conditions. It was observed that the maximum adsorption obtained was between 93% and 98%. under the studies conditions [28]. Wang et al. [29] had reported the use of treated and non-treated fly ash for the removal of methylene blue and basic dye from a wastewater solution. The adsorption capacity for acid treated fly ash was found to be 2.4 x 10-5 mol/g, while non-treated fly ash showed an adsorption capacity of 1.4 x 10-5 mol/g. Wang et al. [30] in another investigation also found that the porous unburned carbon in the fly ash was responsible for the adsorption of the dye, and not the fly ash itself. Table (1) shows selected adsorbent used for dyes removal from polluted water

5. Adsorption of phenols

Since 1860, phenol has been in production, with its basic use as an antiseptic. During late 19th century and thereafter the use of phenol has been further extended to the synthesis of dyes, aspirin, plastics, pharmaceuticals, petrochemical and pesticide chemical industries. In fact, by 2001, the global phenol production has reached an impressive 7.8 million tons [44].

Among the different organic pollutants in wastewater, phenols are considered as priority pollutants since they are harmful to plants, animals and human, even at low concentrations. The major sources of phenolic are steel mills, petroleum refineries, pharmaceuticals,petrochemical, coke oven plants, paints,coal gas, synthetic resins, plywood industries and mine discharge. The wastewater with the highest concentration of phenol (>1000 mg/L) is typically generated from coke processing. Phonolic compounds are also emanated from resin plants with a concentration range of 12–300 mg/L. Environmental Protection Agency (EPA) has set a limit of 0.1 mg/L of phenol in wastewater. The World Health Organization (WHO) is stricter on phenol regulation. It sets a 0.001 mg/L as the limit of phenol concentration in potable water.

Adsorption of phenolic compounds from aqueous solutions by activated carbon is one of the most investigated of all liquid-phase applications of carbon adsorbents [45]. Several adsorbents were used treatment wastewater and removal of phenols. The adsorption isotherms for mono-, di-, and trichlorophenols from aqueous solutions on wood-based and lignite-based carbons were investigated. The adsorptive capacity for 2,4-DCP was found to be 502 mg/g and Freundlich model gave a best fit the experimental data [46]. Zogorski et al. [47] studied the kinetics of adsorption of phenols on GAC. They observed that 60% to 80% of the adsorption occurs within the first hour of contact followed by a very slow approach to the final maximum equilibrium concentration.

In another study, the extent of adsorption of 2,4-dichlorophenol was found to be a function of pH. The presence of surface functional groups also affected the adsorption of phenols onto activated carbon. The presence of dissolved oxygen on activated carbon increased the adsorptive capacity for phenolic compounds This increase in adsorptive capacity was attributed to the oligomerization of the compounds through oxidative coupling reactions [48].

Hamdaouia et al. [49] studied and modeled the adsorption equilibrium isotherms of five phenolic compounds from aqueous solutions onto GAC. The five compounds selected were Phenol (Ph), 2-chlorophenol (2-CP), 4-chlorophenol (4-CP), 2,4-dichlorophenol (DCP), and 2,4,6-trichlorophenol (TCP). They also observed that the interaction of phenolic compounds with activated carbon surface occurred in localized monolayer adsorption type, i.e. adsorbed molecules are adsorbed at definite, localized sites. Uptake of phenols increased in the order Ph < 2-CP < 4- CP < DCP < TCP, which correlated well with respective increase in molecular weight, cross-sectional area, and hydrophobicity and decrease in solubility and pKa.

Sawdust, a very low cost adsorbent was used, after carbonization, for the removal of phenol from industrial waste waters. The equilibrium adsorption level was determined as a function of the solution pH, temperature, contact time, adsorbent dose and the initial concentration. The adsorption maximum for phenol using sawdust was 10.29 mg/L [50].

Adsorbents, carbonaceous materials, activated carbon (AC), bagasse ash (BA) and wood charcoal (WC), were used for removal of phenol from water [51].The results showed the removal efficiencies for phenol–AC, phenol–WC and phenol–BA, approximately 98%, 90% and 90%, respectively. Removal efficiency of phenol slightly increased when the pH of adsorption system decreased. Yapar and Yilmar [52] reported the adsorptive capacity of some clays and natural zeolite materials found in Turkey for the removal of phenol. They found that calcined hydrotalcite was the best among the studied adsorbents in which adsorbed 52% of phenol from a solution of 1000 mg/L phenol at the adsorbent/phenol ratio of 1:100 while the others adsorbed only 8% of phenol. Also, silica gel, activated alumina, AC, fitrasorb 400 and Hisir 1000 adsorbent were examined as adsorbents for the removal of phenol from aqueous solution. They found that Hisir 1000 was the best among the tested materials [53]. Das and Patnaik [54] utilized blast furnace flue dust (BFD) and slag to investigate phenol adsorption through batch experiment.

Bromophenols (2-bromophenol, 4-bromophenol and 2, 4- dibromophenol) considered as one of toxic organic phenol. Industrial wastes was used as low cost adsorbent for the removal of these pollutants. The results show the maximum adsorption on carbonaceous adsorbent prepared from fertilizer industry waste 40.7, 170.4 and 190.2 mg g^{-1} for 4-bromophenol 2-bromophenol and 2, 4-dibromophenol, respectively. As compared to carbonaceous adsorbent, the other three adsorbents (viz., blast furnace sludge, dust, and slag) adsorb bromophenols to a much smaller extent [55]. Table (2) represented the adsorption efficiencies of different adsorbents for the removal of phenols.

Organic Pollutants	Adsorbent	Adsorption Maxmium	Ref.
Phenol	Porous Clay	14.5 mg/g	[56]
2,5-dichlorophenol	Porous Clay	45.5 mg/g	[56]
3,4-dichlorophenol	Porous Clay	48.7 mg/g	[56]
3,5-dichlorophenol	(cetyl-pyridinium-Al PILC),	97.2 mg/g	[57]
phenol	Hemidesmus Indicus Carbon(HIC)	370 ppm	[58]
phenol	Commercial Activated Carbon(CAC)	294 ppm	[58]
phenol	NORIT Granular Activated Carbon (NAC 1240)	74.07 mg/g	[59]
phenol	NORIT Granular Activated Carbon 1010	166.6 mg/g	[59]
phenol	Active carbon	257 mg/g	[60]
phenol	Mesoporous carbon CMK-3-100°C	347 mg/g	[60]
phenol	Mesoporous carbon CMK-3-130 °C	428 mg/g	[60]
phenol	Mesoporous carbon CMK-3-150 °C	473 mg/g	[61]
phenol	Leaf litter of *Shorea roubsta*	76%	[60]
phenol	activated phosphate rock (1M HNO3)	83.34 mg/g	[62]
phenol	Natural clay	15 mg/g	[63]
2,4-dichlorophenol	activated carbon derived from oil palm empty fruit bunch (EFB)	232.56 mg/g	[64]

Table 2. Adsorption capacities of different adsorbents for the removal of phenols.

6. Adsorption of pesticides and herbicides

Pesticides and herbicides, intentionally released into the environment, are ubiquitous in aquatic systems; they are often detected at low levels and commonly occur in the form of complex mixtures [64-65]. Leaching of chemical fertilizers and pesticides, applied to agricultural and forest land, is one of the main reasons for organic pollution in several water streams. Pesticides and herbicides are harmful to life because of their toxicity, carcinogenicity and mutagenicity [66]. Therefore toxicity of pesticides and their degradation products is making these chemical substances a potential hazard by contaminating the environment. They have raised serious concerns about aquatic ecosystem and human health because of the long-term accumulation of their single and/or combined toxicological effects [67]. The contamination of ground water,surface water and soils, by pesticides and herbicides are currently a significant concern, and this because of increasing use of pesticides in agriculture, and domestic activities [68].

Among newly developed pesticides, organophosphorous pesticides are most commonly used. This class of chemicals is divided into several forms; however the two most common forms are phosphates and phosphorothionates. Methyl parathion (O,O-dimethyl O-4-nitrophenyl phosphorothioate) is a class I insecticide. Once methyl parathion introduced into the environment from spraying on crops, droplets of methyl parathion in the air fall on soil, plants or water. While most of the methyl parathion will stay in the areas where it is applied, some can move to areas away from where it was applied by rain, fog and wind [69].

Modified polymer adsorbents were prepared for the removal of organic pollutants from water and wastewater. Adsorption of organic pollutants using cyclodextrin-based polymer (CDPs) as adsorbent, is an efficient technique with the advantages of specific affinity, low cost and simple design [65, 70-71]. Cyclodextrin polymers (CDPs) can be synthesized using cyclodextrin (CD) as complex molecule and polyfunctional substance (e.g., epichlorohydrin (EPI)) as cross-linking agent.Though a number of CDPs with various structures and properties have been developed [72,73], it is still ambiguous how CDP properties affect adsorption affinity toward organic contaminants, particularly mixed pollutants. Liu et al. [66] illustrated the cross-linked structure of cyclodextrin polymer and related adsorption mechanisms in Scheme (1).

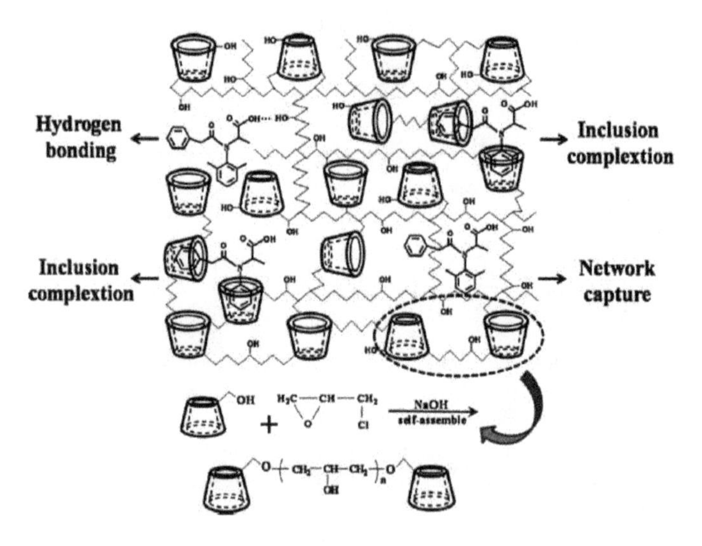

Scheme 1. Cross-linked structure of cyclodextrin polymer and related adsorption mechanisms, from Liu et al. [66].

Other modified polymer adsorbent was used in herbicides treatment, porous polymeric adsorbents were used for the adsorption of herbicides (alachlor, amitrole, trifluralin and prometryn) from liquid solution. Two adsorbent resins were investigated, the highly hydrophobic Amberlite XAD-4 (polystyrene–divinylbenzene copolymer) and the functionalized more hydrophilic XAD-7 (nonionic aliphatic acrylic polymer). The adsorption were sussccufuly at pH 6.5 [74].

Other adsorbent used widely for the removal of pesticides is activated carbon. Activated carbons (ACs) prepared from agricultural and industrial wastes were used for the removal of pesticides from polluted water. Activated carbons produced from agricultural residues (olive kernel, corn cobs, rapeseed stalks and soya stalks) via physical steam activation were tested for the removal of Bromopropylate (BP) pesticide from water. The results show maximum adsorption capacity(q_m) of the pesticide on adsorbents (ACs) prepared from corn cob, olive kernel, soya stalks and rapeseed stalks at values 7.9×10^{-2},12.3×10^{-2}, 11.6×10^{-2} and 18.9×10^{-2} mmol/L, respectively. The BP removal from water achieved in this study was 90–100% for all ACs [75].

Ayranci and Hoda [76] studied the adsorption of pesticides,pesticides ametryn [2- (ethylamino)-4-isopropylamino-6-methyl-thio-s-triazine], aldicarb (2-methyl-2-(methylthio) propionaldehyde o-methylcarbamoyloxime], diuron [N-(3,4-dichlorophenyl)- N,N-dimethyl urea] and dinoseb [2-(sec-butyl)- 4,6-dinitrophenol], from aqueous solution onto high specific area activated carbon-cloth adsorbent (ACC) and found that the maximum adsorption capacity of ACC for ametryn, diuron, dinose and aldicarb were 354.61,421.58, 301.84 and 213.06 mg/g, respectively.

Djilani et al. [77] developed new activated carbon adsorbents from lignocellulosic wastes of vegetable origin (coffee grounds (CG), melon seeds (MS) and orange peels (OP)). The adsorption efficiency of these new adsorbents was tested with organic pollutants: o-nitrophenol and p-nitrotoluene. The elimination ratio obtained with new adsorbents was in the range from 70% to 90%. The time necessary to attain the adsorption equilibrium was between 75 and 135 min.

The ability of MgAl layered double hydroxides (LDHs) and their calcined products to adsorb besticides contaminants, 2,4-dinitrophenol (DNP) and 2-methyl-4,6-dinitrophenol (DNOC) from water was assessed. Adsorption tests were conducted on LDHs with variable Mg/Al ratios (and variable layer charge), pH values, contact times and initial pesticide concentrations to identify the optimum conditions for the intended purpose. All adsorbents except the carbonate-containing hydrotalcite possessed a very high adsorption capacity for both contaminants. As noted above, the adsorption of the pesticides on the calcined LDHs was only 25–40% [78]

Activated carbon prepared from banana stalk by potassium hydroxide (KOH) and carbon dioxide (CO_2) activation (BSAC) was explored for its ability to remove the pesticides, 2,4-dichlorophenoxyacetic acid (2,4-D) and bentazon. The percent removal efficiency of 2,4-D decreased from 98.4 to 85.4% as the 2,4-D initial concentration increased from 50 to 300 mg L^{-1}. For bentazon, the percent removal efficiency decreased from 96.5 to 61.6% as the bentazon initial concentration increased from 25 to 250 mg L^{-1}. Therefore, the adsorption of 2,4-D and bentazon by BSAC has strong dependence on the initial concentration of the pesticides. Maximum adsorption capacity (qe) for 2,4-dichlorophenoxyacetic acid (2,4-D) were 168.03 mg g^{-1}, and for bentazon 100.95 mg g^{-1} [79]

In the other study for using activated carbon as adsorbent, magnetic and graphitic carbon nanostructures was used for the removal of a pesticide (2,4-dichlorophenoxyacetic acid) from aqueous solution.The magnetic and graphitic carbon nanostructures as adsorbents were prepared from two different biomasses, cotton and filter paper. The resultant adsorbents were characterized with TEM and N2 adsorption–desorption methods. The adsorption capacities for the prepared adsorbents from filter paper and cotton are about 77 and 33 mg/g, respectively [80].

Chemically and thermally treated watermelon peels (TWMP) have been utilized for the removal of methyl parathion (MP) pesticide from water. The effect of process variables such as pH of solution, shaking speed, shaking time, adsorbent dose, concentration of solution and temperature have been optimized. Maximum adsorption (99±1%) was achieved for (0.38–3.80)×10−4 mol dm^{-3} of MP solution, using 0.1 g of adsorbent in 20 ml of solution for 60 min agitation time at pH 6. The developed adsorption method has been employed to surface water samples with percent removal 99%±1 [81].

7. Adsorption of other organic pollutants

Other organic pollutants were found as a pollutants in water and wastewater, this include pharmaceutical effluents, surfactants, organic solvents, phthalates, hydrocarbons, esters, alcohols,volatile, semi-volatile and non-volatile chlorinated organic pollutants. Activated carbons, calys and clay minerals are used widely for the removal of organic pollutants.

7.1. Adsorption on activated carbon

Adsorption on activated carbon is currently the most frequently used technology for removing organic pollutants from aqueous industrial sludge, surface waters and drinking water. Methyl tert-butyl ether (MTBE) is an organic pollutants used mainly as a fuel component in fuel if gasoline engine and also as a solvent. The adsorption of methyl tert-butyl ether (MTBE) by granular activated carbon was investigated, the maximum adsorption capacity of MTBE on granular activated carbon was 204.1 mg/g. Results illustrate that granular activated carbon is an effective adsorbent for methyl tert-butyl ether and also provide specific guidance into adsorption of methyl tert-butyl ether on granular activated carbon in contaminated groundwater [82].

A novel triolein-embedded activated carbon composite adsorbent was developed. Results suggested that the novel composite adsorbent was composed of the supporting activated carbon and the surrounding triolein-embedded cellulose acetate membrane. The adsorbent was stable in water, for no triolein leakage was detected after soaking the adsorbent for five weeks. The adsorbent had good adsorption capability to dieldrin, which was indicated by a residual dieldrin concentration of 0.204 μgL^{-1}. The removal efficiency of the composite adsorbent was higher than the traditional activated carbon adsorbent [83]. Essa et al. [84] studied the potential of chemical activated date pits as an adsorbent. The date pits were impregnated with 70% phosphoric acid followed by thermal treatment between 300 to 700°C. The effects of activation temperature and acid concentration on pore surface area development were studied. Samples prepared at 500°C showed a specific area of 1319 m^2/g and total pore volume of 0.785 cm^3/g. Aqueous phenol adsorption trends using the local activated carbon sample were compared to a commercial sample (Filtrasorb-400).

Five commercially available types of activated carbon (GAC 1240, GCN 1240, RB 1, pK 1-3, ROW 0.8 SUPRA)are prepared and used to remove organic chlorinated compounds from wastewater of a chemical plant. The various types of activated carbon were tested on the basis of Freundlich adsorption isotherms for 14 pure organic chlorinated compounds, of molecular weight ranging from that of dichloromethane (MW ¼ 84.93 g mol^{-1}) to hexachlorobenzene (MW ¼ 284.78 g mol^{-1}). The best adsorbent (GAC 1240 granulated activated carbon) was selected and used in a laboratory fixed bed column to assess its removal efficiency with respect to the tested organic chlorinated compounds. Removal efficiency was always higher than 90% (Table 3) [85]

Substance	Precenage of adsorption efficiency (%)
Dichloromethane	98.3
Trichloromethane	98.8
1,1,1- Trichloromethane	99.0
Carbon tetrachloride	99.0
1,2-Dichloroethane	82.8
Trichloroethylene	94.7
1,1,2- Trichloroethane	86.3
Tetrachloroethylene	91.6
1,1,1,2- Tetrachloroethane	87.3
Trans 1,4-dichloro-2-butene	94.2
1,2,4-Trichlorobenzene	99.2
1,2,3-Trichlorobenzene	90.5
Hexachloro-1,3-butadiene	99.4
Hexachlorobenzene	95.1

Table 3. Removal efficiency (%) of chlorinated compounds from wastewater by five commercially available types of activated carbon, from Pavonia et al [85]

Antibiotic considered as one of the pharmacological organic pollutants. Choi et al. [86] investigated the treatment of seven tetracycline classes of antibiotic (TAs) from raw waters (synthetic and river) using coagulation and granular activated carbon (GAC) filtration. Their results ended to that both coagulation and GAC filtration were effective for the removal of TAs, and the removal efficiency depended on the type of TAs. GAC filtration was relatively more effective for removal of tetracycline (TC), doxycycline-hyclate (DXC), and chlortetracycline-HCl (CTC), which were difficult to be removed by coagulation. Putra et al. [87] investigated the removal of Amoxicillin (antibiotic) from pharmaceutical effluents using bentonite and activated carbon as adsorbents. The study was carried out at several pH values. Langmuir and Freundlich models were then employed to correlate the equilibria data on which both models fitted the data equally well. While chemisorption is the dominant adsorption mechanism on the bentonite, both physisorption and chemisorption played an important role for adsorption onto activated carbon. Ruiz et al. [88] studied the removal of paracetamol (anti analgesic drug) from aqueous solutions using chemically modified activated carbons. The effect of the chemical nature of the activated carbon material such as carbon surface chemistry and composition on the removal of paracetamol was studied. The surface heterogeneity of the carbon affected the rate of paracetamol removal. They found that after oxidation the wettability of the carbon was enhanced, which favored the transfer of paracetamol molecules to the carbon pores. At the same time the overall adsorption rate and removal efficiency are reduced in the oxidized carbon due to the competitive effect of water molecules. Tris (2-chloroethyl) phosphate (TCEP), iopromide, naproxen, carbamazepine, and caffeine drugs were quite frequently observed in both surface waters and effluents from waste water treatment plants. The elimination of these chemicals during drinking water and wastewater treatment processes at full- and pilot-scale also was investigated. Conventional drinking water treatment methods such as flocculation and filtration were relatively inefficient for contaminant removal, while efficient removal (99%) was achieved by granular activated carbon (GAC).

Liu et al. [89] studied the removal effects of organic pollutants in drinking water (42 species organic pollutants in 11 categories) by activated carbon, haydite and quartz sand with the method of solid-phase extraction (SPE). The removal rates of total peak area of organic pollutants by activated carbon, haydite and quartz were 70.35%, 29.68% and 37.36%. Among all, activated carbon showed the best removal effect to most organic pollutants contents, and quartz sand to species. So if activated carbon - quartz sand combined processes were adopted, organic pollutants species and total peak area could be reduced simultaneously. The removal rates for phthalates, hydrocarbons, esters and alcohols were 62.62%, 75.83%, 72.52% and 62.99%, respectively. The adsorption capability of activated carbon for dimethyl phthalate and di-n-butyl phthalate, two priority pollutants in water, were preferable. The removal rates reached 93.27% and 57.02%. For haydite, there were 26 kinds of organic pollutants in 10 categories in corresponding treated water. The total peak area was removed by 29.68%. The removal effects for amines, alkanes and phenols were satisfactory and the removal rates were 68.71%, 49.97% and 41.19%, respectively. But in the case of phthalates, esters and aldehydes the effects were not obviously, the same as dimethyl phthalate and di-nbutyl phthalate. The species of organic pollutants were reduced from 36 to 20 by quartz sand. Most notably, xylene, dimethyl phthalate and di-n-butyl phthalate in raw water were removed efficiently. Xylene and di-n-butyl phthalate in tap water were removed absolutely, and dimethyl phthalate was removed by 59.59%.

Cellulose acetate (CA) embedded with triolein (CA-triolein), was prepared as adsorbent for the removal of persistent organic pollutants (POPs) from micro-polluted aqueous solution. The comparison of CA-triolein, CA and granular activated carbon (GAC) for dieldrin removal was investigated. Results showed that CA-triolein absorbent gave a lowest residual concentration after 24 h although GAC had high removal rate in the first 4 h adsorption. Then the removal efficiency of mixed POPs (e.g. aldrin, dieldrin, endrin and heptachlor epoxide), absorption isotherm, absorbent regeneration and initial column experiments of CA-triolein were studied in detail. The linear absorption isotherm and the independent absorption in binary isotherm indicated that the selected POPs are mainly absorbed onto CA-triolein absorbent by a partition mechanism. Thermodynamic calculations showed that the absorption was spontaneous, with a high affinity and the absorption was an endothermic reaction. Rinsing with hexane the CA-triolein absorbent can be regenerated after absorption of POPs. No significant decrease in the dieldrin removal efficiency was observed even when the absorption–regeneration process was repeated for five times. The results of initial column experiments showed that the CA-triolein absorbent did not reach the breakthrough point at a breakthrough empty-bed volume (BV) of 3200 when the influent concentration was 1–1.5 l g/L and the empty-bed contact time (EBCT) was 20 min [90]

Activated coke (AC) was studied to adsorb organic pollutants from coking wastewater. The study initially focused on the sorption kinetics and equilibrium sorption isotherms of AC for the removal of chemical oxygen demand (COD) from coking wastewater. The results showed that when the dose of AC was 200 g L^{-1}, 91.6% of COD and 90% of color could be removed after 6 h of agitation at 40°C. The kinetics of adsorption of COD from coking wastewater onto AC was fit to the pseudo-second order model. The adsorption of COD onto AC was enhanced with an increase of temperature, indicating that the adsorption process would be a chemical adsorption rather than a physical one [91].

Bottom ash, a kind of waste material generated from thermal coal-fired power plants, is generally used in road bases and building materials. Bottom ash was used to remove the organic pollutants in coking wastewater and papermaking wastewater. Particular attention was paid on the effect of bottom ash particle size and dosage on the removal of chemical oxygen demand (COD). The results show that the COD removal efficiencies increase with decreasing particle sizes of bottom ash, and the COD removal efficiency for coking wastewater is much higher than that for papermaking wastewater due to its high percentage of particle organic carbon (POC). Different trends of COD removal efficiency with bottom ash dosage are also observed for coking and papermaking wastewaters because of their various POC concentrations [92].

7.2. Adsorption on clays and clay minerals adsorbents

Several adsorbents were used for the removal of these pollutants. One of an effective and low cost adsorbents is calys and clay minerals. Natural clay minerals due to their high surface area and molecular sieve structure are very effective sorbents for organic contaminants of cationic or polar in character.Natural and modified clay minerals and zeolites are good candidates for improving activated carbon(AC) performance, because they have large surface areas for retention of pollutants [93].

Adsorbent prepared from organoclays and activated carbon were shown to remove a variety of organic contaminants [94-95]. Montmorillonite was applied as adsorbent for the removal of cationic surfactants, while the hydrophilic surface of montmorillonite was modified and used as adsorbent [96]. Calcined hydrocalcites was prepared and used to remove organic anionic pesticides [97] from polluted water. Vesicle–clay complexes in which positively charged vesicles composed of didodecyldimethy- lammonium bromide (DDAB) were adsorbed on montmorillonite and removed efficiently anionic (sulfentrazone, imazaquin) and neutral (alachlor, atrazine) pollutants from water. These complexes (0.5% w:w) removed 92–100% of sulfentrazone, imazaquin and alachlor,and 60% of atrazine from a solution containing 10 mg/L. A synergistic effect on the adsorption of atrazine was observed when all pollutants were present simultaneously (30 mg/L each), its percentage of removal being 85.5. Column filters (18 cm) filled with a mixture of quartz sand and vesicle–clay (100:1, w:w) were tested. For the passage of 1 L (25 pore volumes) of a solution including all the pollutants at 10 mg/L each, removal was complete for sulfentrazone and imazaquin, 94% for alachlor and 53.1% for atrazine, whereas removal was significantly less efficient when using activated carbon. A similar advantage of the vesicle–clay filter was observed for the capacities of removal. Table (4) show the removal efficiencies of organic pollutants from water by DDAB-Clay complex adsorbent [98]

Data in parentheses correspond to the procedure when dried complex was added to the solution, whereas the other case correspond to the removal after incubation with DDAB followed by clay addition.

Data in brackets are the predicted removal by Langmuir equation (binding coeffients)

Zhao et al. [99] prepared mesoporous silica materials and used it for adsorption of organic pollutants in water. Mesoporous silica materials is performed using self-assembling micellar aggregates of two surfactants: cetylpyridinium bromide (CPB) and cetyltrimethylammonium bromide (CTAB). The retention properties have been studied of these two kinds mesoporous silicas towards environmental pollutants (mono-, di-, tri-chloroacetic acid, toluene, naphthalene and methyl orange). The effect of the composition (presence and absence of surfactants, different kinds of surfactants) on the sorption performance has been considered. They found that materials show excellent retention performance toward chloroacetic acids, toluene, naphthalene and methyl orange. The materials without surfactants does not show, if any, affinity for ionic and non-ionic analytes.

The applicability of mesoporous aluminosilica monoliths with three-dimensional structures and aluminum contents with 19≤Si/Al≥1 was studied as effective adsorbents of organic molecules from an aqueous solution. Mesocage cubic Pm3n aluminosilica monoliths were successfully fabricated using a simple, reproducible, and direct synthesis (scheme 2). The acidity of the monoliths significantly increased with increasing amounts of aluminum species in the silica pore framework walls. The batch adsorption of the organic pollutants onto (10 g/L) aluminosilica monoliths was performed in an aqueous solution at various temperatures. These adsorbents exhibit efficient removal of organic pollutants (e.g., aniline, pchloroaniline,o-aminophenol, and p-nitroaniline) of up to 90% within a short period (in the order of minutes). In terms of proximity adsorption, the functional acid sites and the condensed and rigid monoliths with tunable periodic scaffolds of the cubic mesocages are useful in providing easy-to-use removal assays for organic compounds and reusable adsorbents without any mesostructural damage, even under chemical treatment for a number of repeated cycles [100].

% removal	Initial herbicide conc. (mg/L)	DDAB Conc.(Mm)	Pollutants
37.3±0.3 (57.4± 0.7) 56.6±1.0 (59.8±0.1) [59.0]	8 8	3 5	Atrazine
94.9±0.8 (85.6±1.0) 95.5±0.1 (87.2±0.2) [92.7]	10 10	3 6	Alachlor
73.6±0.4 (92.0±0.3) 75.52±0.3 (92.3±1.0) [95.8]	10 10	3 6	Imazaquin
98.0±1.0 (99.7±0.1) 99.7±0.1 (99.7±0.1) [99.9]	10 10	3 6	Sulfentrazone

Table 4. Removal of organic pollutants from water by DDAB-Clay complex adsorbent [a,b] (From Undabeytiaa et al. [98])

Modified clays were used as adsorbents for the removal of organic pollutants from wastewater. Two pillared clays are synthesized by intercalation of solutions of aluminium and zirconium and evaluated as adsorbents for the removal of Orange II and Methylene Blue from aqueous solutions. The contact time to attain equilibrium for maximum adsorption was found to be 300 min. Both clays were found to have the same adsorption capacity when Orange II was used as adsorbent, whereas the adsorption capacity of Zr-PILC was higher (27 mg/g) than that of Al-PILC (21 mg/g)for Methylene Blue. The adsorption kinetics of dyes has been studied in terms of pseudo-first- and -second-order kinetics, and the Freundlich, Langmuir and Sips isotherm models have also been applied to the equilibrium adsorption data. The addition of NaCl has been found to increase the adsorption capacities of the two pillared clays for Orange II Pillared InterLayered Clays (PILCs) are porous materials that can be obtained by the intercalation of soils, thereby creating high value added materials from natural solids [101]. Also, pillared clay adsorbent was used for the removal of benzo(a)pyrene and chlorophenols [102], chlorinated phenols from aqueous solution by surfactant-modified pillared clays [103] and herbicide Diuron on pillared clays [104].

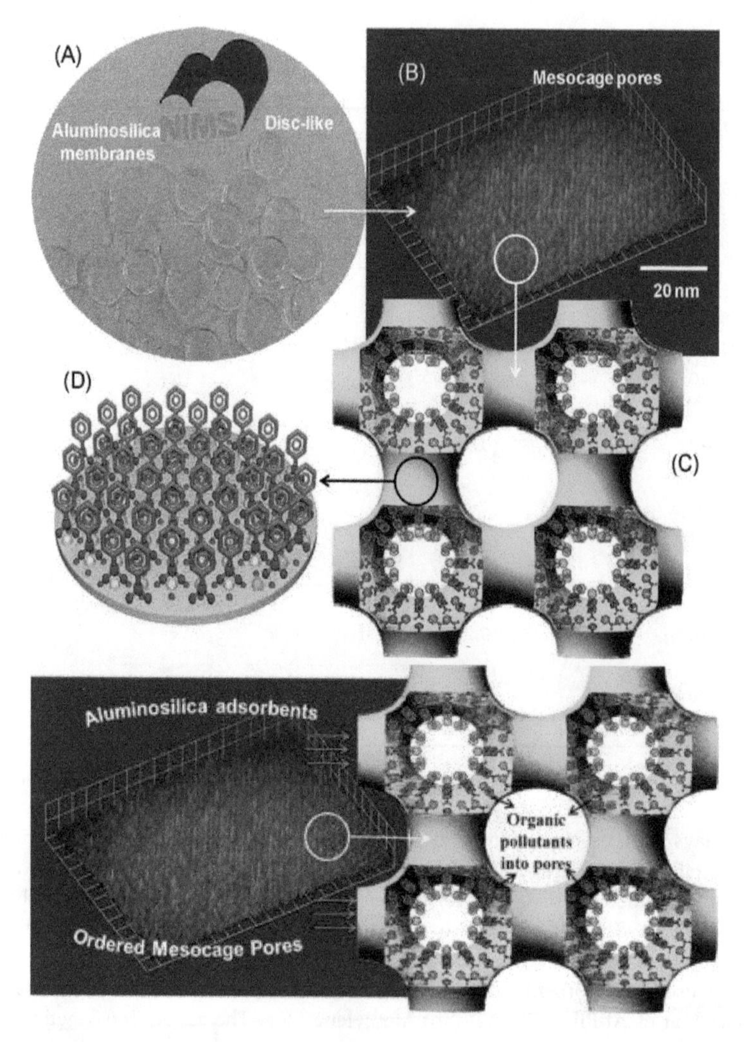

Scheme 2. Aluminosilica monoliths with a disc-like shape (A) and mesocage pores (B) as adsorbents (C) of organic compounds (I–IV) inside the mesocage cavity and onto pore surfaces of 3D cubic Pm3n structures (D). Note that 3D TEM image (B) was recorded with aluminosilica monoliths with a Si/Al ratio of 4 (From El-Safty et al. [100]).

8. Conclusion

Organic pollutants in the ecosystem, especially persistent organic pollutants (POPs),are of the most important environmental problems in the world. The literature reviewed revealed that there has been a high increase in production and utilization of organic pollutants in last few decades resulting in a big threat of pollution. Efficient techniques for the removal of highly toxic organic compounds from water and wastewater have drawn significant interest. Adsorption is recognized as an effective and low cost technique for the removal of organic pollutants from water and wastewater, and produce high-quality treated effluent. This chapter highlighted the removal of organic pollutants using adsorption technique with different kinds of natural and synthetic adsorbents.

Many researches have given considerable attention aimed at establishing to the removal efficiency of organic pollutants by adsorption technique. To decrease treatment costs, attempts have been made to find inexpensive alternative activated carbon (AC), from waste materials of industrial, domestic and agricultural activities. Also, clays and natural clay minerals, due to their high surface area and molecular sieve structure, are very effective adsorbents for organic contaminants. The chapter focus, reviews and evaluates literature dedicated on the adsorption phenomenon, different types of natural and synthetic adsorbents, adsorption of dyes, phenols, pesticides and other organic pollutants. Finally it ended with recent researches of organic pollutants adsorption on activated carbons, clays and clay minerals.

Abbreviations

AC Activated carbon, PCP Pentachlorophenol

CAC Commercial Activated carbon, PCDD/Fs polychlorinated dibenzo-p-dioxins and dibenzofurans

PAC powdered activated carbons, DDT dichloro-diphenyl-trichloroethane

ACC Activated carbon cloth adsorbent, DNOC 2,4-dinitrophenol (DNP) and 2-methyl-4,6-dinitrophenol

WC Wood charcoal, HCB Hexachlorobenzene

GAC Granular activated carbon, CD Cyclodextrin

HIC Hemidesmus Indicus carbon, BP Bromopropylate

POPs Persistent organic pollutants, LDHs MgAl layered double hydroxides

BET Buruner, Emmett, and Teller, MTBE Methyl tert-butyl ether

TCP 2,4,6-trichlorophenol, MP Methyl parathion

DCP Dichlorophenol, TCEP Tris (2-chloroethyl) phosphate

BA Bagasse ash, TC Tetracycline

BFD Blast furnace flue dust, DXC Doxycycline-hyclate

TWMP Treated watermelon peels, CTC Chlortetracycline-HCl

PPC Potato peels charcoal, CA Cellulose acetate

WBA Wood-shaving bottom ash, CA-triolein Cellulose acetate (CA) embedded with triolein

CDPs Cyclodextrin-based polymer, EBC Empty-bed contact time

CG Coffee grounds, DDAB Ddidodecyldimethy- lammonium bromide

MS Melon seeds, CPB Cetylpyridinium bromide

OP Orange peels, CTAB Cetyltrimethylammonium bromide

MB Methylene blue, EPA Environmental Protection Agency

RR141 Red Reactive, 141 WHO TheWorld Health Organisation

POPs Persistent organic pollutants, PhAcS Pharmaceutically active substances

OCPs Organochlorine pesticides, PCBs polychlorinated biphenyls

TMP Trimethoprim, PCP Pentachlorophenol

PhAcS Pharmaceutically active substances, PCDD/Fs polychlorinated dibenzo-p-dioxins and dibenzofurans

PCBs polychlorinated biphenyls, DDT Dichloro-diphenyl-trichloroethane

COD Chemical oxygen demand, MW Molecular Weight

Author details

Mohamed Nageeb Rashed

Address all correspondence to: mnrashed@hotmail.com

Aswan Faculty of Science, Aswan University, Aswan, Egypt

References

[1] Ali, I., Mohd. Asim, Tabrez A. Khan. Low cost adsorbents for the removal of organic pollutants from wastewater. Journal of Environmental Management 2012;113,170-183.

[2] Ali, I., Aboul-Enein, H.Y., Chiral. Pollutants: Distribution, Toxicity and Analysis by Chromatography and Capillary Electrophoresis. John Wiley & Sons, Chichester, 2004.

[3] Damià, B. Emerging Organic Pollutants in Waste Waters and Sludge. Springer, Berlin, 2005.

[4] Harrad, S., Persistent Organic Pollutants Enviromental Behaivour and Pathways for Human Exposure. Kluwer Academic Publishers, Norwell,2001.

[5] Burkhard, L.P., Lukazewycz, M.T. Toxicity equivalency values for polychlorinated biphenyl mixtures. Environ. Toxicol. Chem. 2008, 27, 529–534.

[6] Clive, T. Persistent Organic Pollutants: Are we close to a solution? Published by the Canadian Arctic Resources Committee,Volume 26, Number 1, Fall/Winter.2000.

[7] Fiedler, H..Persistent Organic Pollutants. Springer, New York, 2003.

[8] Ritter, L. ; Solomon, K.R.; Forget, J. Persistent organic pollutants. An assessment report Canadian Network of Toxicology Centres 620 Gordon Street Guelphon Canada, 2000.

[9] Crini G. Recent developments in polysaccharide-based materials used as adsorbents in wastewater treatment,.Prog. Polym Sci. 2005,30, 38–70,

[10] Kurniawan, T. A. and Lo, W.H.. Removal of refractory compounds from stabilized landfill leachate using an integrated H2O2 oxidation and granular activated carbon (GAC) adsorption treatment. Water Research,2009, 43, 4079-4091.

[11] Aly, Osman M and Faust, Samuel D. Adsorption processes for water treatment. Butterworth Publisher (Boston) 509 p, 1987.

[12] Moreno-Castilla C. Adsorption of organic molecules from aqueous solutions on carbon materials. Carbon 2004, 42,83–94.

[13] Amokrane, A., Comel, C. and Veron, J. Landfill leachate pretreatment by coagulation-flocculation. Water Research,1997, 31, 2775-2782.

[14] [14] Surmacz-Gorska, J. Degradation of Organic Compounds in Municipal Landfill Leachate. Lublin, Publishers of Environmental Engineering Committee of Polish Academy of Sciences,2001.

[15] Putraa, E. K., Pranowo, R., Sunarsob, J., Indraswati, N. and Ismadji, S. Performance of activated carbon and bentonite for adsorption of amoxicillin from wastewater: Mechanisms, isotherms and kinetics water research. 2009, 43, 2419 - 2430.

[16] Lillo-Rodenas M.A., D. Cazorla-Amoros, A. Linares-Solano. Behaviour of activated carbons with different pore size distributions and surface oxygen groups for benzeneand toluene adsorption at low concentrations. Carbon 2005, 43,1758–1767.

[17] Ardejani F. Doulati, Kh. Badii, N. Yousefi Limaee, N.M. Mahmoodi, M. Arami, S.Z. Shafaei, A.R. Mirhabibi.Numerical modelling and laboratory studies on the removal

of Direct Red 23 and Direct Red 80 dyes from textile effluents using orange peel, a low-cost adsorbent. Dyes and Pigments 2007,73, 178-185.

[18] Gupta Vinod K., Alok Mittal, Rajeev Jain, Megha Mathur, Shalini S. Adsorption of Safranin-T from wastewater using waste materials- activated carbon and activated rice husks Journal of Colloid and Interface Science,2006,303, 80–86.

[19] Hamdaoui Oualid. Batch study of liquid-phase adsorption of methylene blue using cedar sawdust and crushed brick Journal of Hazardous Materials 2006,B135,264–273.

[20] Leechart P., Woranan Nakbanpote b,1, Paitip Thiravetyan. Application of 'waste' wood-shaving bottom ash for adsorption of azo reactive dye Journal of Environmental Management 2009,90,912–920

[21] Tsai Wen-Tien, Hsin-Chieh Hsub, Ting-Yi Su b, Keng-Yu Lin b, Chien-Ming Lin. Removal of basic dye (methylene blue) from wastewaters utilizing beer brewery waste. Journal of Hazardous Materials 2007,154 (1-3)73-78.

[22] Rashed. M.N.. Acid Dye Removal from Industrial Wastewater by Adsorption on Treated Sewage Sludge, Int. J. Environment and Waste Management, 2011, 7, Nos. 1/2,175-191.

[23] Rozada F., M. Otero, A. Moran, A.I. Garcıa. Adsorption of heavy metals onto sewage sludge-derived materials. Bioresource Technology 2009,99, 6332–6338.

[24] Rio, S., Catherine, F.B., Laurence, L.C., Philippe, C. and Pierre, L.C. Experimental design methodology for the preparation of carbonaceous sorbents from sewage sludge by chemical activation-application to air and water treatments. Chemosphere 2005,58, 423–437.

[25] Martin, M.J., Artola, A., Balaguer, M.D. and Rigola, M..Towards waste minimization in WWTP: activated carbon from biological sludge and its application in liquid phase adsorption, J. Chem. Technol. Biotechnol., 2002,77, 825–833.

[26] Gulnaz, O., Kayaa, A., Matyar, F. and Arikan, B. Sorption of basic dyes from aqueous solution by activated sludge', Journal of Hazardous Materials B, 2004, 108, 183–188.

[27] Otero M., F. Rozada, L.F. Calvo, A.I. Garcıa, A. Moran Adsorption on adsorbents from sewage sludge Elimination of organic water pollutants using adsorbents obtained from sewage sludge Dyes and Pigments 57,2003, 55–65.

[28] Acemioglu, 2004Acemioglu,B. Adsorption of Congo red from aqueous solution onto calcium rich fly ash. Journal of Colloid Interface Science, 2004, 274, 371-379.

[29] Wang, S., Boyjoo, Y., Choueib, A. and Zhu, J. Utilization of fly ash as low cost adsorbents for dye removal. Chemeca, 2004,26-29.

[30] Wang, S., Boyjoo, Y., and Zhu, J. Sonochemical treatment of fly ash for dye removal from wastewater. Journals of Hazardous Materials, 2005,Vol. B126, 91-95.

[31] Al-Degs Yahya S. a, Musa I. El-Barghouthi a, Amjad H. El-Sheikh a, Gavin M. Walker. Effect of solution pH, ionic strength, and temperature on adsorption behavior of reactive dyes on activated carbon, Dyes and Pigments 2008,77, 16-23.

[32] Orhan Ozdemira, Bulent Armaganb, Mustafa Turanb, Mehmet S. C¸ elikc. Comparison of the adsorption characteristics of azo-reactive dyes on mezoporous minerals,Dyes and Pigments 2004, 62, 49–60.

[33] Indra D. Mall, Vimal C. Srivastava, Nitin K. Agarwal. Removal of Orange-G and Methyl Violet dyes by adsorption onto bagasse fly ashdkinetic study and equilibrium isotherm analyses, Dyes and Pigments 2006,69,210-223

[34] Mansoor Anbia, Samira Salehi. Removal of acid dyes from aqueous media by adsorption onto amino-functionalized nanoporous silica SBA-3, Dyes and Pigments 2012, 94,1-9

[35] Auta M., B.H. Hameed. Preparation of waste tea activated carbon using potassium acetate as an activating agent for adsorption of Acid Blue 25 dye, Chemical Engineering Journal 2011, 171, 502– 509

[36] Emad N. El Qada, Stephen J. Allen, Gavin M. Walker. Adsorption of Methylene Blue onto activated carbon produced from steam activated bituminous coal: A study of equilibrium adsorption isotherm. Chemical Engineering Journal 2006, 124, 103–110

[37] G. Gong, Q. Xie, Y. Zheng, S. Ye, Y. Chen. Regulation of pore size distribution in coal-based activated carbon, New Carbon Mater. 2009, 24, 141–146

[38] B.S. Girgis, E. Smith, M.M. Louis, A.N.A. El-Hendawy. Pilot production of activated carbon from cotton stalks using H3PO4, J. Anal. Appl. Pyrol. 2009,86, 180–184

[39] Mehmet Ulas Durala, Levent Cavasa,b, Sergios K. Papageorgiouc, Fotis K. Katsarosc. Methylene blue adsorption on activated carbon prepared from Posidonia oceanica (L.) dead leaves: Kinetics and equilibrium studies. Chemical Engineering Journal 2011, 168, 77–85

[40] Yongze Bao, Guilan Zhang. Study of Adsorption Characteristics of Methylene Blue onto Activated Carbon Made by Salix Psammophila. Energy Procidia 2012,16, 1141 – 1146

[41] Alexandro M.M. Vargas, Andre L. Cazetta, Marcos H. Kunita, Tais L. Silva, Vitor C. Almeida. Adsorption of methylene blue on activated carbon produced from flamboyant pods (Delonix regia): Study of adsorption isotherms and kinetic models. Chemical Engineering Journal 2011, 168,722–730

[42] Ahmad A.L., M.M. Loh, J.A. Aziz. Preparation and characterization of activated carbon from oil palm wood and its evaluation on methylene blue adsorption, Dyes Pigments 2007, 75, 263–272.

[43] Tan I.A.W., A.L. Ahmad, B.H. Hameed. Enhancement of basic dye adsorption uptake from aqueous solutions using chemically modified oil palm shell activated carbon, Colloids Surf. A 2008,318, 88–96.

[44] Das V.Srihari– Ashutosh. Adsorption Of Phenol From Aqueous Media By An Agro-Waste (Hemidesmus Indicus) Based Activated Carbon.Applied Ecology And Environmental Research 2011,7(1): 13-23.

[45] Radovic, L. R., Moreno-Castilla, C. and Rivera-Utrilla, J. Carbon materials as adsorbents in aqueous solutions. In: Radovic, L.R. (Ed.). Chemistry and Physics of Carbon, A Series of Advances, 2001,27, 227-405.

[46] Dobbs R. A. and Cohen, J. M.. Carbon adsorption isotherms for toxic organics. In: EPA-600/8-80-023, Municipal Environmental Research Laboratory, Office of Research and Development, Cincinnati, Ohio 1980.

[47] Zogorski, J. S., Faust, S. D. and Haas Jr., J. H. The kinetics of adsorption of phenols by granular activated carbon. J. Colloid Interface Sci.,1976, 55, 329-341.

[48] Snoeyink, V. L., McCreary, J. J. and Murin, C. J. Carbon adsorption of trace organic compounds. Cincinnati, Ohio, Municipal Research Lab Office of Research and Development, U.S. Environmental Protection Agency,1977.

[49] Hamdaoui, O. and Naffrechoux, E. Modeling of adsorption isotherms of phenol and chlorophenols onto granular activated carbon Part I. Two-parameter models and equations allowing determination of thermodynamic parameters. Journal of Hazardous Materials,2007, 147, 381-394.

[50] Larous. S1, Meniai A-H1 M.F.F. Sze, V.K.C. Lee, G. McKay.The use of sawdust as by product adsorbent of organic pollutant from wastewater: adsorption of phenol. Desalination 2008,218, 323–333

[51] Mukherjee Somnath, Sunil Kumarb, Amal K. Misra, Maohong Fan.Removal of phenols from water environment by activated carbon, bagasse ash and wood charcoal. Chemical Engineering Journal 2007,129, 133–142.

[52] Yapar S., M. Yilmar. Removal of phenol by using montmorillonite, clinoptilolite and hydrotalcite, Adsorption 2004, 10, 287–298.

[53] Roostari N., F.H. Tezel. Removal of phenol from aqueous solution by adsorption, J. Environ. Manage. 2004,70, 157–164.

[54] Das C.P., L.N. Patnaik. Removal of phenol by industrial solid waste, Practice Period. Hazard. Toxic Radioactive Waste Manage. ASCE 9,2005, (2) 135–140.

[55] Bhatnagar Amit. Removal of bromophenols from water using industrial wastes as low cost adsorbents. Journal of Hazardous Materials 2007,B139, 93–102.

[56] Sofía Arellano-Cárdenas,Tzayhrí Gallardo-Velázquez, Guillermo Osorio-Revilla, Ma. del Socorro López-Cortéz1 and Brenda Gómez-Perea. Adsorption of Phenol and Di-

chlorophenols from Aqueous Solutions by Porous Clay Heterostructure (PCH) J. Mex. Chem. Soc. 2005, 49(3), 287-291

[57] Matthes, W., Kahr, G. Sorption of organic compounds by Al- and Zr-hydroxy-intercalated and pillared bentonite. Clays and Clay Minerals 2000,48, 593–602.

[58] V.Srihari, Ashutosh Das.Adsorption Of Phenol From Aqueous Media By An Agro-Waste (Hemidesmus Indicus) Based Activated Carbon. Applied Ecology And Environmental Research 2009, 7(1): 13-23.

[59] Hawaiah Imam Maarof, Bassim H. Hameed, Abdul Latif Ahmad. Adsorption Isotherms for phenols onto Activated Carbon. AJChE,2004,4(1)-70-76

[60] Enamul Haque, Nazmul Abedin Khan, Siddulu Naidu Talapaneni,Ajayan Vinu, Jonggeon Jegal and Sung Hwa Jhung. Adsorption of Phenol on Mesoporous Carbon CMK-3: Effect of Textural Properties, Bull. Korean Chem. Soc. 2010, Vol. 31, No. 6

[61] Mishra S., Bahattachrya J. Potential of leaf litter for phenol adsorption-A Kinetic Study. Indian J.Chem Techn. 2006,13, 298-301

[62] Atef S. Alzaydien and Waleed Manasreh. Equilibrium, kinetic and thermodynamic studies on the adsorption of phenol onto activated phosphate rock. International Journal of Physical Sciences 2009, 4 (4), 172-181,

[63] Djebbar M. 1, F. Djafri1, M. Bouchekara and A. Djafri. Adsorption of phenol on natural clay.African Journal of Pure and Applied Chemistry 2012, 6(2), 15-25, 15.

[64] Shaarani F.W., B.H. Hameed.Batch adsorption of 2,4-dichlorophenol onto activated carbon derived from agricultural waste, Desalination, 2010,255(1–3)159–164

[65] Gilliom, R.J. Pesticides in U.S. streams and groundwater. Environmental Science & Technology 2007,41 (10), 3408-3414.

[66] Liu Huihui, Xiyun Cai, Yu Wanga, Jingwen C. Adsorption mechanism-based screening of cyclodextrin polymers for adsorption and separation of pesticides from water.Water research 2011,4 5, 3 4 9 9 -3 5 1 1

[67] International Agency for Research on Cancer (IARC), Overall evaluations of carcinogenicity: An updating of IARC monographs volumes 1 to 42, Supplement 7, WHO, Lyon, France, 1987.

[68] Cui, N., Zhang, X., Xie, Q.,Wang, S.,Chen, J.,Huang, L., Qiao, X., Li, X., Cai, X. Toxicity profile of labile preservative bronopol in water: the role of more persistent and toxic transformation products. Environmental Pollution 2011,159 (2), 609-615.

[69] Philip H.H., in: E.M. Michalenko,W.F. Jarvis, D.K. Basu,G.W. Sage,W.M. Meyland, J.A. Beauman, D.A. Gray (Eds.), Handbook of Environmental Fate and Exposure Data for Organic Chemicals, III, Lewis, Chelsea, MI, 1991.

[70] Memon G. Zuhra, M.I. Bhanger, Mubeena Akhtar, Farah N. Talpur, Jamil R. Memon. Adsorption of methyl parathion pesticide from water using watermelon peels as a low cost adsorbent.Chemical Engineering Journal, 2008,138, 616–621

[71] Allabashi, R., Arkas, M., Hormann, G., Tsiourvas, D. Removal of some organic pollutants in water employing ceramic membranes impregnated with cross-linked silylated dendritic and cyclodextrin polymers. Water Research 2007,41 (2), 476-486

[72] Romo, A., Penas, F.J., Isasi, J.R., Garcia-Zubiri, H.X., Gonzalez- Gaitano, G. Extraction of phenols from aqueous solutions by beta-cyclodextrin polymers. Comparison of sorptive capacities with other sorbents. Reactive & Functional Polymers 2008, 68 (1), 406-413.

[73] Crini, G., Morcellet, M. Synthesis and applications of adsorbents containing cyclodextrins. Journal of Separation Science 2002, 25 (13), 789-813.

[74] G. Kyriakopoulosa, D. Douliaa, E. Anagnostopoulos. Adsorption of pesticides on porous polymeric adsorbents.Chemical Engineering Science 2005,60, 1177 – 1186

[75] Ioannidou Ourania A., Anastasia A. Zabaniotou, George G. Stavropoulos, Md. Azharul Islam, Triantafyllos A. Albanis. Preparation of activated carbons from agricultural residues for pesticide adsorption. Chemosphere 2010,80,1328–1336

[76] Ayranci Erol, Numan Hoda. Adsorption kinetics and isotherms of pesticides onto activated carbon-cloth. Chemosphere 2005,60, 1600–1607

[77] Djilani Chahrazed, Rachida Zaghdoudib, Ali Modarressid, Marek Rogalskid, Fayc¸al Djazia, Abdelaziz Lallame. Chemical Engineering Journal 2012,189– 190, 203– 212

[78] Chaara D., I. Pavlovic, F. Bruna, M.A. Ulibarri, K. Draoui, C. Barrig. Removal of nitrophenol pesticides from aqueous solutions by layered double hydroxides and their calcined products. Applied Clay Science 2010,50, 292–298

[79] SalmanJ.M., V.O. Njokua, B.H. Hameeda. Adsorption of pesticides from aqueous solution onto banana stalk activated carbon. Chemical Engineering Journal 2011,174, 41– 48

[80] Maryam Khoshnood, Saeid Azizian.Adsorption of 2,4-dichlorophenoxyacetic acid pesticide by graphitic carbon nanostructures prepared from biomasses.Journal of Industrial and Engineering Chemistry.2012,18(5) 1796-1800

[81] Memon G. Zuhra, M.I. Bhanger, Mubeena Akhtar, Farah N. Talpur, Jamil R. Memon. Adsorption of methyl parathion pesticide from water using watermelon peels as a low cost adsorbent..Chemical Engineering Journal 2008,138, 616–621

[82] Chen D. Z.,J. X. Zhang,J. M. Chen. Adsorption of methyl tert-butyl ether using granular activated carbon: Equilibrium and kinetic analysis Int. J. Environ. Sci. Tech.,2010, 7 (2), 235-242.

[83] Jia RU, LIU Huijuan, QU Jiuhui, WANG Aimin, DAI Ruihua.Characterization and adsorption behavior of a novel triolein-embedded activated carbon composite adsorbent.Chinese Science Bulletin 2005,50 (23)2788—2790.

[84] Essa, M.H.;Al-Zahrani M.A.; Nesaratnam S.A. Step towards national reliance using locally produced activated carbon from dates. Third Saudi Technical Conference and Exhibition (STCEX-3), Riyadh, Saudi Arabia, 291-297,2004.

[85] Pavonia B., D. Drusiana, A. Giacomettia, M. Zanetteb.Assessment of organic chlorinated compound removal from aqueous matrices by adsorption on activated carbon. Water Research 2006,40, 3571 – 3579.

[86] Choi, K. J., Kima, S. G. and Kim, S. H. Removal of antibiotics by coagulation and granular activated carbon filtration. Journal of Hazardous Materials, 2008,151(1), 38-43.

[87] Putraa, E. K., Pranowo, R., Sunarsob, J., Indraswati, N. and Ismadji, S. Performance of activated carbon and bentonite for adsorption of amoxicillin from wastewater: Mechanisms, isotherms and kinetics. Water research. 2009,43, 2419 – 2430.

[88] Ruiz, B., Cabrita, I., Mestre, A. S., Parra, J. B., PiresJ., Carvalho, A. P. and Ania, C. O. Surface heterogeneity effects of activated carbons on the kinetics of paracetamol removal from aqueous solution. Applied Surface Science, 2010,256, 5171-5175.

[89] Liu Hanchao, Feng Suping, Du Xiaolin, Zhang Nannan, Liu Yongli. Comparison of three sorbents for organic pollutant removal in drinking water. Energy Procedia 2012,18, 905 – 914.

[90] Huijuan Liu, Jia Ru, Jiuhui Qu, Ruihua Dai, Zijian Wang, Chun Hu.Removal of persistent organic pollutants from micro-polluted drinking water by triolein embedded absorbent. Bioresource Technology 2009,100,2995–3002

[91] Mo he Zhang, Quan lin Zhao, Xue Baib, Zheng fang Ye. Adsorption of organic pollutants from coking wastewater by activated coke Colloids and Surfaces A: Physicochem. Eng. Aspects 2010,362, 140–146

[92] Wei-ling Sun, Yan-zhi Qu, Qing Yu, Jin-Ren Ni. Adsorption of organic pollutants from coking and papermaking wastewaters by bottom ash. Journal of Hazardous Materials 2008,154, 595–601

[93] Watanabe, Y., Yamada, H., Tanaka, J., Komatsu, Y., Moriyoshi, Y. Environmental purification materials: removal of ammonium and phosphate ions in water system. Trans. Mater. Res. Soc. Jpn. 2004, 29, 2309–2312.

[94] Beall, G.W. The use of organo-clays in water treatment. Appl. Clay Sci.2003, 24, 11–20.

[95] Ake, C.L., Wiles, M.C., Huebner, H.J., McDonald, T.J., Cosgriff, D.,Richardson, M.B., Donelly, K.C., Phillips, T.D. Porous organoclay composite for the sorption of polycy-

clic aromatic hydrocarbons and pentachlorophenol from groundwater. Chemosphere 2003,51, 835–844.

[96] Borisover, M., Graber, E.R., Bercovich, F., Gerstl, Z. Suitability of dye–clay complexes for removal of non-ionic organic compounds from aqueous solutions. Chemosphere 2001,44, 1033–1040.

[97] Li, F., Wang, Y., Yang, Q., Evans, D.G., Forano, C., Duan, X. Study on adsorption of glyphosate (N-phosphonomethyl glycine) pesticide on MgAl-layered double hydroxides in aqueous solution. J. Hazard. Mater. 2005, 125, 89–95.

[98] Tomas Undabeytiaa,, Shlomo Nirb, Trinidad Sanchez-Verdejoa, Jaime Villaverdea, Celia Maquedaa, Esmeralda Morilloa. A clay–vesicle system for water purification from organic pollutants.Water Research 2008,42, 1211 – 1219.

[99] Zhao Y.X., M.Y. Ding, D.P. Chen. Adsorption properties of mesoporous silicas for organic pollutants in water. Analytica Chimica Acta 2005,542,193–198.

[100] Sherif A. El-Safty, Ahmed Shahata, Mohamed Ismaela. Mesoporous aluminosilica monoliths for the adsorptive removal of small organic pollutants. Journal of Hazardous Materials 2012,201- 202, 23- 32.

[101] Gil A., F.C.C. Assis, S. Albeniz, S.A. Korili Removal of dyes from wastewaters by adsorption on pillared clays. Chemical Engineering Journal 2011,168,1032–1040.

[102] Srinivasan K.R., H.S. Fogler. Use of inorgano-organo-clays in the removal of priority pollutants from industrial wastewaters: adsorption of benzo(a)pyrene and chlorophenols from aqueous solutions, Clays Clay Miner.1990, 38, 287–293.

[103] Michot L.J., T.J. Pinnavaia. Adsorption of chlorinated phenols from aqueous solution by surfactant-modified pillared clays, Clays Clay Miner. 1991,39, 634–641.

[104] Bouras O., J..C. Bollinger, M. Baudu, H. Khalaf. Adsorption of diuron and its degradation products from aqueous solution by surfactant-modified pillared clays, Appl. Clay Sci. 2007,37, 240–250.

Chapter 8

Application of Different Advanced Oxidation Processes for the Degradation of Organic Pollutants

Amilcar Machulek Jr., Silvio C. Oliveira,
Marly E. Osugi, Valdir S. Ferreira, Frank H. Quina,
Renato F. Dantas, Samuel L. Oliveira,
Gleison A. Casagrande, Fauze J. Anaissi,
Volnir O. Silva, Rodrigo P. Cavalcante, Fabio Gozzi,
Dayana D. Ramos, Ana P.P. da Rosa, Ana P.F. Santos,
Douclasse C. de Castro and Jéssica A. Nogueira

Additional information is available at the end of the chapter

1. Introduction

Water is not only an economic, but also an increasingly important social commodity. Potable water is an essential resource for sustaining economic and social development in all sectors. A safe water supply and appropriate sanitation are the most essential components for a healthy and prosperous life. However, increases in human activities have led to exposure of the aqueous environment to chemical, microbial and biological pollutants as well as to micro-pollutants. Thus, liquid effluents containing toxic substances are generated by a variety of chemistry-related industrial processes, as well as by a number of common household or agricultural applications.

New, economically viable, more effective methods for pollution control and prevention are required for environmental protection and effluent discharge into the environment must have minimal impact on human health, natural resources and the biosphere.

Research in photochemical and photocatalytic technology is very promising for the development of viable alternatives for the treatment of polluted waters and effluents from various sources, including both industrial and domestic. Currently available chemical and photochemical technology permits the conversion of organic pollutants with a wide range of chemical structures into substances that are less toxic and/or more readily biodegradable by employing chemical oxidizing agents in the presence of an appropriate catalyst and/or ultraviolet light to oxidize or degrade the pollutant of interest. These technologies known as advanced oxidation processes (AOP) or advanced oxidation technologies (AOT), have been widely studied for the degradation of diverse types of industrial wastewaters. These processes are particularly interesting for the treatment of effluents containing highly toxic organic compounds, for which biological processes may not be applicable unless bacteria that are adapted to live in toxic media are available. The production of powerful oxidizing agents, such as the hydroxyl radical, is the main objective of most AOP. The hydroxyl radical reacts rapidly and relatively non-selectively with organic compounds by hydrogen abstraction, by addition to unsaturated bonds and aromatic rings, or by electron transfer. In the case of persistent organic pollutants (wastes), complete decontamination may require the sequential application of several different decontamination technologies such as a pretreatment with a photochemical AOP followed by a biological or electrochemical treatment.

This chapter discusses the influence of different AOP on the degradation and mineralization of several different classes of organic pollutants such as pesticides, pharmaceutical formulations and dyes. The use of the Fenton and photo-Fenton reactions as tools for the treatment of pesticides and antineoplastic agents is presented, as well as examples of the optimization of the important parameters involved in the process such as the source of iron ions (free or complexed), the irradiation source (including the possibility of using sunlight), and the concentrations of iron ions and hydrogen peroxide. The chapter also reports the use of TiO_2 nanotubes obtained by electrochemical anodization, nanoparticles prepared by a molten salt technique, and Ag-doped TiO_2 nanoparticles as heterogeneous photocatalysts, emphasizing their potential for use in environmental applications. These catalysts were characterized by a combination of techniques, including scanning electron microscopy, elemental analysis, and energy dispersive x-ray spectroscopy.

2. Advanced Oxidation Processes (AOP)

AOP are specific chemical reactions characterized by the generation of chemical oxidizing agents capable of oxidizing or degrading the pollutant of interest. The efficiency of the AOP is generally maximized by the use of an appropriate catalyst and/or ultraviolet light [1-3].

In most AOP, the objective is to use systems that produce the hydroxyl radical ($HO^{\bullet}$) or another species of similar reactivity such as sulfate radical anion ($SO_4^{\bullet-}$). These radicals react with the majority of organic substances at rates often approaching the diffusion-controlled limit (unit reaction efficiency per encounter). Both of these species are thus highly reactive and only modestly selective in their capacity to degrade toxic organic compounds present in aqueous solution. The principal reaction pathways of $HO^{\bullet}$ with organic compounds include hydrogen abstraction from aliphatic carbon, addition to double bonds and aromatic rings, and electron transfer [4]. These reactions generate organic radicals as transient intermediates, which then undergo further reactions, eventually resulting in final products corresponding to the net oxidative degradation of the starting molecule [5].

The AOP are of two main types: homogeneous and heterogeneous processes, both of which can be conducted with or without the use of UV radiation. Thus, for example, the homogeneous process based on the reaction of Fe^{2+} with H_2O_2, known as the thermal-Fenton reaction process typically becomes more efficient for the mineralization of organic material present in the effluent when it is photocatalysed. This latter process (Fe^{2+}/Fe^{3+}, H_2O_2, UV-Vis) is commonly referred to as the photo-Fenton reaction. Among the heterogeneous AOP, processes using some form of the semiconductor TiO_2 stand out because UV irradiation of TiO_2 results in the generation of hydroxyl radicals, promoting the oxidation of organic species [1,6].

2.1. Advances in research on AOP

AOP and their applications have attracted the attention of both the scientific community and of corporations interested in their commercialization. This can be illustrated by means of searches, in August, 2012, of the Science Finder Scholar database (version 2012). This database covers the complete text of articles/papers indexed from over 15475 international journals and 126 databases with abstracts of documents in all areas, as well as several other important sources of academic information. The results of the searches were organized as histograms to show the evolution of the number of publications (articles or patents) related to the different kinds of AOP. Figure 1 shows the results of a search using the keywords "advanced oxidation processes", which yielded approximately 840 publications and which nicely reflects the rapid growth in interest AOP, given the unique characteristics and the versatility of application of AOP.

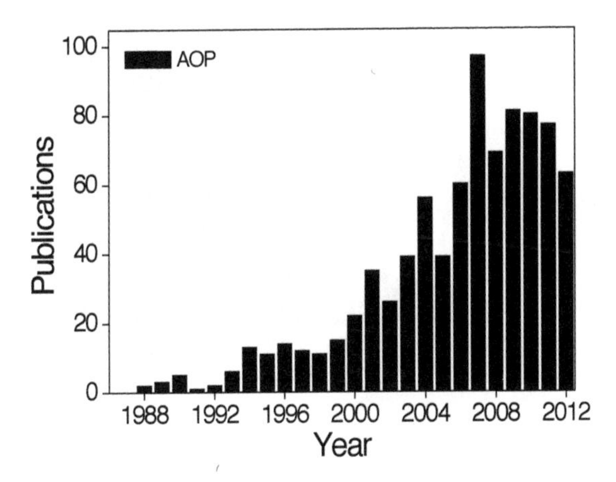

Figure 1. Number of publications per year indexed in the Science Finder Scholar database retrieved using the keywords "advanced oxidation processes".

2.2. Fenton reaction

The thermal Fenton reaction is chemically very efficient for the removal of organic pollutants. The overall reaction is a simple redox reaction in which Fe(II) is oxidized to Fe(III) and H_2O_2 is reduced to the hydroxide ion plus the hydroxyl radical.

$$Fe^{2+} + H_2O_2 \rightarrow Fe^{3+} + HO^{\bullet} + OH^- \quad (1)$$

The ferric ion produced in Equation 1 can in principle be reduced back to ferrous ion by a second molecule of hydrogen peroxide:

$$Fe^{3+} + H_2O_2 \rightarrow Fe^{2+} + HO_2^{\bullet} + H^+ \quad (2)$$

However, this thermal reduction (Equation 2) is much slower than the initial step (Equation 1) and the addition of relatively large, essentially stoichiometric amounts of Fe(II) may be required in order to degrade the pollutant of interest [7]. Another important limitation of the Fenton reaction is the formation of recalcitrant intermediates that can inhibit complete mineralization. Despite these potential limitations, the conventional Fenton reaction has been widely used for the treatment of effluents [6, 8-11].

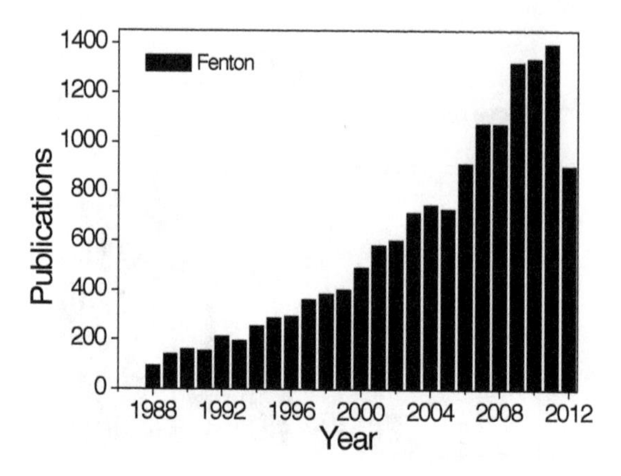

Figure 2. Number of publications per year indexed in the Science Finder Scholar database retrieved using the keyword "Fenton".

For the degradation of organic molecules, the optimum pH for the Fenton reaction is typically in the range of pH 3-4 and the optimum mass ratio of catalyst (as iron) to hydrogen peroxide is 1:5 respectively [12]. One way of accelerating the Fenton reaction is via the addition of catalysts, in general from certain classes of organic molecules such as benzoquinones or dihydroxybenzene (DHB) derivatives [13,14]. It is also possible to accelerate the Fenton reaction via irradiation with ultraviolet light, a process generally known as the photo-assisted Fenton or photo-Fenton reaction, which is discussed in the following section.

Regarding the Fenton reaction, a search of Science Finder Scholar (2012) with the keyword "Fenton" without any refinements retrieved 14821 publications from 1986 to 2012. As shown in Figure 2, there is a clear upward trend in the publications, with 902 publications related to this topic being reported in just the first half of 2012.

2.3. Photo-fenton reaction

One of the most efficient AOP is the photo-Fenton reaction (Fe^{2+}/Fe^{3+}, H_2O_2, UV light), which successfully oxidizes a wide range of organic and inorganic compounds. The irradiation of Fenton reaction systems with UV/Vis light (250-400 nm) strongly accelerates the rate of degradation. This behavior is due principally to the photochemical reduction of Fe(III) back to Fe(II), for which the overall process can be written as:

$$Fe^{3+} + H_2O + hv \rightarrow Fe^{2+} + HO^{\bullet} + H^{+} \quad (3)$$

Studies of the pH dependence of the photo-Fenton reaction have shown that the optimum pH range is ca. pH 3. Studies of the photochemistry of $Fe(OH)^{2+}$, which is the predominant species in solution at this pH and that is formed by deprotonation of hexaaquairon(III), have shown that $Fe(OH)^{2+}$ undergoes a relatively efficient photoreaction upon excitation with UV light to produce Fe(II) and the hydroxyl radical. Therefore, irradiation of Fenton reaction systems not only regenerates Fe(II), the crucial catalytic species in the Fenton reaction, but also produces an additional hydroxyl radical, the species responsible for provoking the degradation of organic material. As a consequence of these two effects, the photo-Fenton process is faster than the conventional thermal Fenton process.

The efficiency of the photo-Fenton process can be further enhanced by using certain organic acids to complex Fe(III). Thus, for example, oxalic acid forms species such as $[Fe(C_2O_4)]^{+}$, which absorbs light as far out as 570 nm, i.e., well into the visible region of the spectrum. This species makes the photo-Fenton reaction more efficient because it absorbs a much broader range of wavelengths of light and because, upon irradiation, it efficiently decomposes (quantum yield of the order of unity) to Fe(II) and CO_2:

$$2[Fe(C_2 0_4)]^{+} + hv \rightarrow 2Fe^{2+} + 2CO_2 + C_2O_4^{2-} \quad (4)$$

The use of photo-Fenton reaction has considerable advantages in practical applications. It generally produces oxidation products of low toxic, requires only small quantities of iron salt (which can be either Fe^{3+} or Fe^{2+}) and offers the possibility of using solar radiation as the source of light in the reaction process sunlight constitutes an inexpensive, environmentally friendly, renewable source of ultraviolet photons for use in photochemical processes.

The disadvantages of the photo-Fenton process include the low pH values required and the need for removal of the iron catalyst after the reaction has terminated. If necessary, however, the residual Fe(III) can usually be precipitated as iron hydroxide by increasing the pH. Any residual hydrogen peroxide that is not consumed in the process will spontaneously decompose into water and molecular oxygen, being thus a "clean" reagent itself. These features make homogeneous photo-Fenton based AOPs the leading candidate for cost-efficient, environmental friendly treatment of industrial effluents on a small to moderate scale [6, 15-17]. Currently much research activity is focused on attempts to develop new catalysts that function at neutral pH that do not require acidification of the effluent in order to react and that also do not require removal of the catalyst at the end of the reaction.

A search of Science Finder Scholar (2012) with the keyword "photo-Fenton" (Figure 3) showed a modest increase during the 1990s followed by a much more robust upward trend since ca. 2000.

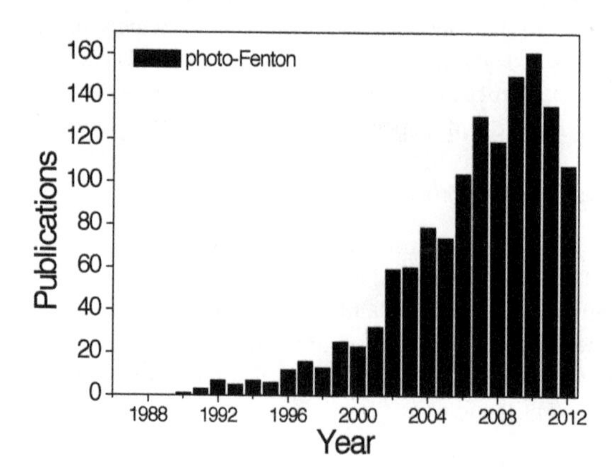

Figure 3. Number of publications per year indexed in the Science Finder Scholar database retrieved using the keyword "photo-Fenton".

2.4. Ozone

Ozone is a powerful oxidizing agent with a high reduction potential (2.07V) that can react with many organic substrates [18,19]. Using ozone, the oxidation of the organic matrix can occur via either direct or indirect routes [20,21]. In the direct oxidation route, ozone molecules can react directly with other organic or inorganic molecules via electrophilic addition. The electrophilic attack of ozone occurs on atoms with a negative charge (N, P, O, or nucleophilic carbons) or on carbon-carbon, carbon-nitrogen and nitrogen-nitrogen pi-bonds [22,23]. Indirectly, ozone can react via radical pathways (mainly involving HO•) initiated by the decomposition of ozone.

A process that employs ozone is only characterized as an AOP when the ozone decomposes to generate hydroxyl radicals (Equation 5), a reaction that is catalyzed by hydroxide ions (OH^-) in alkaline medium or by transition metal cations [18,24,25].

$$2O_3 + 2H_2O \rightarrow 2HO^{\bullet} + O_2 + 2HO_2^{\bullet} \tag{5}$$

The efficiency of ozone in degrading organic compounds is improved when combined with H_2O_2, UV radiation or ultrasound. The initial step in the UV photolysis of ozone is dissociation to molecular oxygen and an oxygen atom (Equation 6), which then reacts with water to produce H_2O_2 (Equation 7):

$$O_3 + hv \rightarrow O_2 + O^{\bullet} \tag{6}$$

$$H_2O + O^{\bullet} \rightarrow H_2O_2 \tag{7}$$

In a second photochemical step (Equation 8), H_2O_2 photodissociates into the active species, two hydroxyl radicals:

$$H_2O_2 + hv \rightarrow 2HO^{\bullet} \tag{8}$$

The O_3/UV process has been employed commercially to treat ground water contaminated with chlorinated hydrocarbons, but cannot compete economically with the H_2O_2/UV process. A major problem with the use of ozone for water treatment is bromine formation in waters containing bromide ion. Strategies such as addition of H_2O_2 (O_3/H_2O_2) can reduce bromine formation and assure the suitability of ozone for treating drinking and wastewater [26].

A search of the Science Finder Scholar database retrieved using only the keyword "ozone" retrieved, as expected, an enormous number of publications, nearly 130,000. Refinement with the additional keyword "degradation" reduced this to 3057 publications, which is a significant number when compared with other AOPs, especially in recent years (Figure 4).

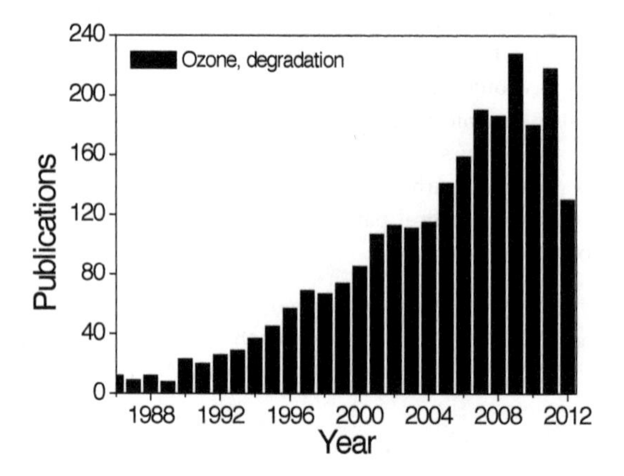

Figure 4. Number of publications per year indexed in the Science Finder Scholar database retrieved using the keywords "ozone" and "degradation".

2.5. Heterogeneous AOP

Another important class of AOP is based on the use of solid semiconductors as heterogeneous catalysts for the mineralization of organic compounds. In this type of photocatalysis, an electron in the valence band of the semiconductor (CdS, TiO_2, ZnO, WO_3, etc.) is promoted into the conduction band upon excitation. The electron in the conduction band typically reacts with O_2, while the hole in the valence band can react with an adsorbed pollutant or oxidize water to produce a surface-bound $HO^\bullet$ radical [2].

According to Alfano and coworkers [27], the anatase form of titanium dioxide (TiO_2) is the material most indicated for use in photocatalytic water treatment, considering aspects such as toxicity, resistance to photocorrosion, availability, catalytic efficiency and cost. Using TiO_2 as the semiconductor, the photocatalysis is based on the activation of anatase by light [28]. The band gap or energy difference between the valence and conduction bands of anatase is 3.2 eV. Thus, UV light of wavelength shorter than 390 nm is capable of exciting an electron (e^-) from the valence to the conduction band.

$$TiO_2 + hv \rightarrow e^- + h^+ \tag{9}$$

An important feature of TiO_2 photocatalysis is the very high oxidation potential of the holes left in the valence band (3.1 eV at pH 0), making it possible for photoexcited TiO_2 to oxidize most organic molecules.

The electron (e^-) and hole (h^+) pair produced by absorption of UV light can migrate to the surface of the anatase particle, where they react with adsorbed oxygen, water, hydroxide ion or organic species via electron transfer reactions. Both water and hydroxide ion can act as electron donors to the holes (h^+) of the catalyst [27,29], generating hydroxyl radicals, as shown by Equations 10 and 11.

$$h^+ + H_2O \rightarrow HO^\bullet + H^+ \tag{10}$$

$$h^+ + HO^- \rightarrow HO^\bullet \tag{11}$$

When dissolved molecular oxygen is present or is deliberately added to the medium, it can act as an acceptor of the electron in the conduction band, generating the superoxide radical (Equation 12) and triggering a series of reactions that can lead to the formation of hydroxyl radicals [30,31].

Alternatively, one can increase the oxidative efficiency of TiO_2 photocatalysis by adding H_2O_2. The electrons in the conduction band then reduce the added H_2O_2 to $HO^\bullet$ and HO^- [32], according to Equation 13.

$$O_2 + e^- \rightarrow O_2^- \tag{12}$$

$$H_2O_2 + e^- \rightarrow HO^\bullet + OH^- \tag{13}$$

The use of TiO_2 also makes it possible to degrade organic molecules that are resistant to oxidation, since they can potentially be reduced by the electrons in the conduction band.

TiO_2 photocatalysis has a number of important advantages in relation to other AOP and, in some aspects, even some biological treatments. In particular, unlike other AOP, the TiO_2/UV system can be employed to treat pollutants in the gas phase, as well as in solution. In addition, TiO_2 has a relatively low cost, is essentially insoluble in water and biologically and chemically inert. Moreover, it can be used to treat effluents containing a wide range of concentrations of pollutants, in particular very low concentrations. Solar radiation can be used to activate the catalyst; and the excellent mineralization efficiency is observed for organochlorine compounds, chlorophenols, nitrogen-containing pesticides, aromatic hydrocarbons, dioxins, carboxylic acids, etc. The principle limitations of TiO_2 photocatalysis in practical applications are the low quantum efficiency of the process and the limited depth of penetration of the incident radiation into suspensions of TiO_2, due to the strong scattering of light by the opaque white catalyst particles. Incrustation of the reactor walls with catalyst can also reduce the amount of incident light. Batch reactors also require additional unit operations in order to physically separate the catalyst from the solution at the end of the irradiation for recycling. Although substantial progress has been made in developing larger-scale reactors for carrying out heterogeneous photochemical reactions, much work remains to be done before TiO_2 photocatalysis becomes a generally applicable technique.

Figure 5 shows the evolution of publications related to heterogeneous photocatalysis by TiO_2, reflecting the potential for application of this technology on an industrial scale.

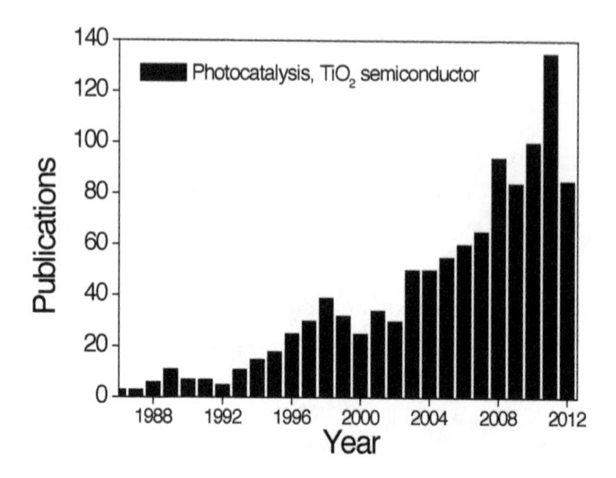

Figure 5. Number of publications per year indexed in the Science Finder Scholar database retrieved using the keywords "photocatalysis" and "TiO_2 semiconductor".

3. Applications

In this section, several applications of homogeneous and heterogeneous AOP are discussed, focusing on the degradation and mineralization of organic pollutants such as pesticides, pharmaceutical formulations and dyes.

3.1. Homogeneous AOP applied to degradation of the herbicide chlorimurom-ethyl and the antineoplastic agent mitoxantrone

The use of the thermal Fenton and the photo-Fenton reactions for the treatment of the pesticide chlorimurom-ethyl (CE) and the antineoplastic agent mitoxantrone (MTX) is described here, along with the optimization of the parameters involved in these processes, including the sources of iron (free or complexed) and irradiation (lamp or possibility of using sunlight) and the concentrations of iron and hydrogen peroxide, etc. Ozone and ozone combined with UV and H_2O_2 were also used as alternative treatments of these pesticides.

3.1.1. *Degradation of Chlorimurom-Ethyl (CE)*

The thermal Fenton, photo-Fenton and ozonation processes were applied for the degradation of a commercial preparation of chlorimurom-ethyl (CE, Figure 6), a compound belonging to the class of sulfonylurea herbicides. This herbicide, widely used in the cultivation of soybeans, may persist in the environment and has residual phytotoxicity [33].

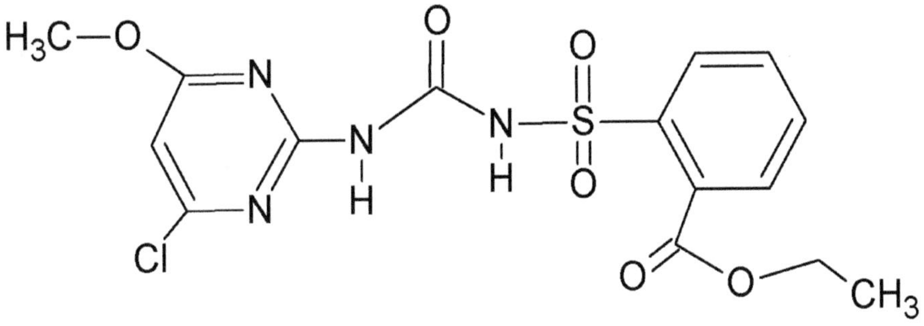

Figure 6. Molecular structure of Chlorimurom-ethyl (CE).

Experiments were performed in a photochemical reactor (1.0 L) equipped with a high pressure mercury lamp (125 W) coupled to a reservoir (2.0 L) via a recirculation pump. The photo-Fenton degradation was influenced by the initial concentrations of H_2O_2 and Fe^{2+}. Experiments were performed with different H_2O_2 concentrations, ranging from 17 to 103 mmol L^{-1}, maintaining the Fe^{2+} concentration constant at 0.33 mmol L^{-1}. Subsequently, the H_2O_2 concentration was fixed at 68.4 mmol L^{-1}, the value that gave the best mineralization, and the Fe^{2+} concentrations were varied from 0.20 to 1.0 mmol L^{-1}. The extent of mineralization of the organic material, expressed as the percentage of removal of the total organic carbon (TOC), ranged from 84% to 95%. Since the quantity of Fe^{2+} had only a small effect on CE removal, a concentration of Fe^{2+} of 0.20 mmol L^{-1} was used in subsequent experiments. In all cases, the extent of mineralization was higher than the percentage of degradation of CE (82-87%) determined by HPLC. This particularity reflects the fact that a commercial formulation of CE was employed in the experiments. Thus, a solution of this formulation in water that contained 30 mg L^{-1} of CE contained 65 mg L^{-1} of total organic carbon. Therefore, it can be concluded that the other organic compounds present in the composition react somewhat better with $HO^{\bullet}$ than CE.

The effect of UV radiation on this optimized reaction system was used to compare the efficiencies of the thermal Fenton and photo-Fenton reactions for the mineralization of CE (Figure 7) with each other and with those of several other homogeneous AOP. Under direct photolysis there was no significant mineralization. Less than 20% TOC removal was obtained at the end of the thermal Fenton treatment. However, a considerable increase in mineralization was observed when the Fenton system was irradiated with UV light. Monitoring CE removal rather than TOC showed that both the thermal Fenton reaction and the photo-Fenton reactions caused extensive degradation of the target compound. Therefore, in the photo-Fenton process, UV radiation makes a significant contribution to both mineralization and CE removal.

Normative Instruction nº 2, published on January 3, 2008, by the Brazilian Ministry of Agriculture (MAPA) [34], regulates the practice in Brazil for treatment of pesticide residues in effluents generated by agricultural aviation companies. The Ministry recommends ozonation for a minimum of six hours using a system with a minimum capacity for producing one gram of ozone per hour for each charge of four hundred and fifty liters of pesticide residue derived from washing and cleaning of aircraft equipment [34]. To verify the efficiency of this system for the mineralization of CE-contaminated water, the samples were treated with ozone alone and with ozone in combination with UV light and H_2O_2. Although oxidation of CE was very fast with all the ozonation methods studied, the use of ozone alone proved to be of limited utility with regard to the mineralization of the organic content of CE-contaminated waters. The combination of $O_3/UV/H_2O_2$ did, however, achieve a high extent of mineralization (80%), indicating that the mineralization of the organic content is mediated by the $HO^{\bullet}$ radical.

When compared to the other systems studied, the photo-Fenton system showed the best results, with mineralization exceeding 85%, making it the preferred technique for the treatment of wastewater containing this pesticide.

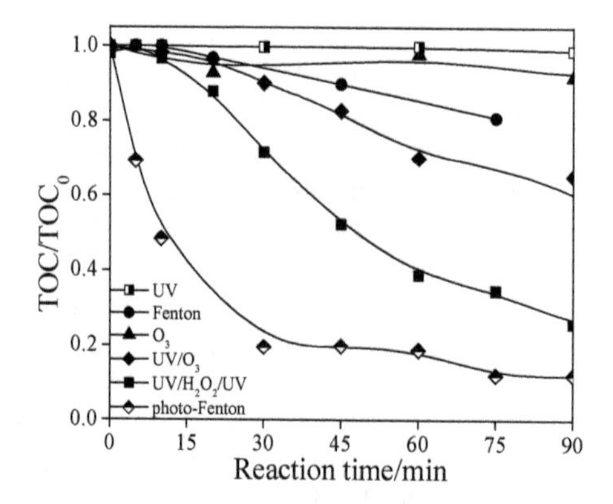

Figure 7. Comparison of the efficiencies of mineralization of a commercial formulation of CE in aqueous solution by the different AOP. $[CE]_0 = 0.060$ mmol L^{-1}; $[TOC]_0 = 65$ mg L^{-1}; when present, $[Fe^{2+}]=0.2$ mmol L^{-1}, $[H_2O_2] = 68.4$ mmol L^{-1} and $[O_3]= 25$ mg mL^{-1}.

3.1.2. Degradation of Mitoxantrone (MTX)

Antineoplastic agents (drugs employed in cancer chemotherapy) are pollutants due their mutagenic, carcinogenic, and genotoxic potential, even at trace levels [35]. The AOP selected for degradation of the antineoplastic drug mitoxantrone (MTX), Figure 8 [36], were the photo-Fenton (with Fe^{2+}, Fe^{3+}, and potassium ferrioxalate - $K_3(FeOx)$ - as iron sources), solar photo-Fenton, Fenton and UV/H_2O_2 reactions. The MTX degradation experiments were carried out using an annular glass photochemical reactor (working volume, 1 L) and a quartz tube for introduction of the radiation source (a 125 W mercury vapor lamp).

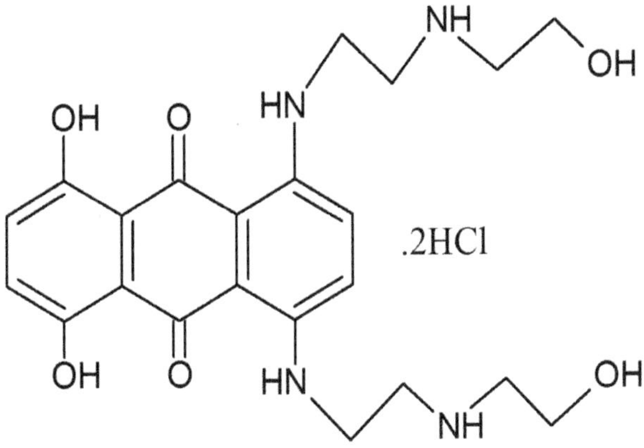

Figure 8. Molecular structure of Mitoxantrone.

Degradation of MTX by the photo-Fenton process was investigated with several different concentrations of Fe(II) (0.54, 0.27, and 0.13 mmol L^{-1}) and H_2O_2 (4.0, 9.4, and 18.8 mmol L^{-1}). The results showed a low removal of TOC, with a mineralization of only 14-35%. One explanation for this low efficiency is that MTX has nitrogen and oxygen atoms that might serve as complexation sites for iron(III), making it unavailable for participation in the Fenton reaction. The possibility of complexation between MTX and iron(III) was investigated by spectrophotometric measurements. Indeed, addition of $Fe(NO_3)_3$ to solutions of MTX caused significant spectral changes, including a shift and a decrease in the absorbance of the long-wavelength absorption band (608-658 nm) of the drug. Spectrophotometric titrations suggested that the complex has a 2:1 Fe^{3+}:MTX stoichiometric ratio with a complexation constant (K) of 1.47×10^4 $M^{-2,}$ indicative of a high affinity of MTX for Fe^{3+}.

In order to minimize the effects of the complexation of Fe(III) by MTX, the use of more stable, but photoactive iron complexes as the source of iron in the degradation process was examined. One such complex is potassium ferrioxalate K_3(FeOx). This complex is often employed because of its high quantum efficiency of photodecomposition and strong absorption in the UV-visible region (up to 500 nm), compatible with the use of solar irradiation in a K_3(FeOx) - mediated photo-Fenton process [37].

Figure 9 compares the efficiencies of several different AOP for the degradation of MTX. The photo-Fenton process employing K_3(FeOx) and the UV/H_2O_2 process were the most efficient for mineralizing MTX, with 82% and 90% total organic carbon removal, respectively. Total degradation of MTX was observed in the thermal Fenton process, but only 65% degradation of MTX occurred under UV irradiation alone; However, TOC data show that there was no appreciable mineralization of MTX under direct photolysis and in the thermal Fenton reaction, even after long treatment periods, whereas the photo-Fenton reaction using solar irradiation led to a TOC removal of 59%.

Although the UV/H_2O_2 process is usually slower than the photo-Fenton process, due to the complexation of MTX with Fe(III) in the latter, the UV/H_2O_2 process proved to be more efficient in this case. To corroborate this, the amount of photogenerated Fe(II) was quantified during the irradiation of ferric ions [Fe(NO3)3] and ferrioxalate in the presence of MTX. In the presence of MTX, the photoreduction of Fe(III) generated only 75 $\mu mol\ L^{-1}$ of Fe(II), while irradiation of ferrioxalate generated 285 $\mu mol\ L^{-1}$ of Fe(II) under the same experimental conditions. This conclusively shows that MTX inhibits the photochemical step of the photo-Fenton reaction, making the overall process substantially less efficient.

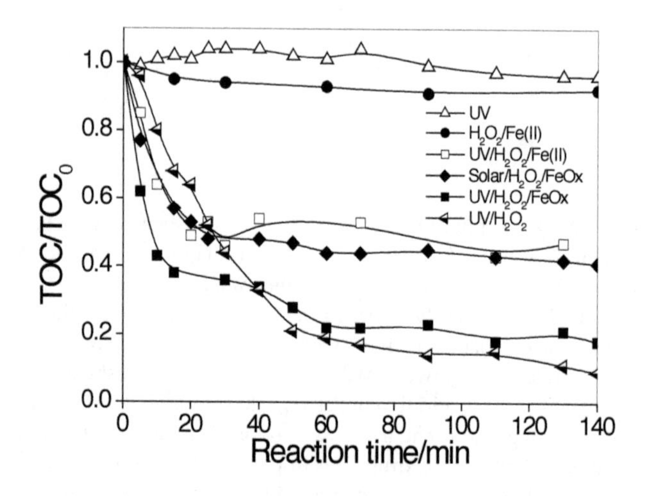

Figure 9. Comparison of the mineralization of aqueous MTX solutions (0.077 mmol L^{-1}) by different AOP (0.54 mmol L^{-1} iron source and 18.8 mmol L^{-1} H_2O_2, when present).

Cytotoxicity evaluation of the solution during treatment by an AOP is a very important since the intermediates and by-products formed during the oxidation of the organic material can be more toxic than the initial target compound. Cytotoxicity tests were performed using NIH/3T3 mouse embryonic fibroblast cells. The concentration (IC_{50}) for inhibition of growth by MTX was 3.29 $\mu g\ mL^{-1}$, demonstrating its toxicity to NIH/3T3 cells. In contrast, 100% growth of NIH/3T3 cells was observed in similar tests on aliquots of solutions of MTX that had been degraded by the H_2O_2/UV and photo-Fenton (UV/H_2O_2/ K_3(FeOx)) processes, indicating an absence of toxic effects. Thus, these two AOP, which degraded MTX completely and exhibited the best mineralizations of the drug, generated no toxic by-products, confirming the potential of both of these processes for the removal of MTX from aqueous solution.

3.2. Preparation of TiO_2 semiconductors and their application in the heterogeneous photocatalysis of methyl viologen, methylene blue and xylidine

TiO_2 is an important, widely studied photocatalytic material [38]. Several samples of TiO_2 are commercially available, but Evonik (Degussa) P-25 (70% anatase and 30% rutile) is the most popular and, in most cases, gives the best results. However, different methods such as sol-gel process [39-41], electrochemical anodization [42], and molten-salt synthesis [43] can be used to prepare TiO_2 in the form of powders, nanoparticles, thin film, nanotubes, etc. This section considers heterogeneous photocatalysis employing TiO_2 in the forms of nanotubes obtained by electrochemical anodization, of nanoparticles prepared by sol-gel or molten salt techniques, and of Ag-doped TiO_2 nanoparticles. These catalysts were characterized by a series of techniques, including scanning electron microscopy, elemental analysis, energy dispersive x-ray spectroscopy, etc. and were applied for the degradation of a herbicide and a dye.

TiO_2 prepared by the sol-gel process (acid hydrolysis of titanium(IV) isopropoxide) was used for the photocatalytic degradation of the herbicide methyl viologen (MV^{2+}, Figure 10), which is widely employed in over 130 countries on crops of rice, coffee, sugar cane, beans, and soybeans, among others [44], despite a high power of intoxication. The performance under irradiation of nanoparticles of TiO_2 prepared by the sol-gel technique (TiO_2 SG) was compared to TiO_2 SG doped with Ag (0.5%-4.0%), and to undoped and doped TiO_2 P25. The materials were characterized by thermogravimetric analysis, X-ray diffraction, surface area, infrared spectroscopy, scanning electron microscopy and energy dispersive spectroscopy. X-Ray diffraction analysis showed that TiO_2 synthesized by the sol-gel method is similar to TiO_2 P25 with both anatase and rutile peaks, but with a lower crystallinity and an increase in the surface area compared to P25. The surface area of TiO_2 SG, determined experimentally by BET, was 71.21 $m^2\ g^{-1}$, 1.5 times larger than TiO_2 P25 (46.18 $m^2\ g^{-1}$). The doping with Ag influenced the values of the band gap energy (E_{gap}), determined by diffuse reflectance spectroscopy. Higher percentages of Ag resulted in a decrease the E_{gap} value, shifting the light absorption to the visible region. Additionally, energy dispersive spectroscopic analysis confirmed the presence of Ag in the doped materials. Scanning electron microscopic (SEM) analysis (Figure 11) indicated that silver changed the oxide morphology, depending on the amount. In materials with 0.5% (Figure 11 A) and 1.0% of Ag (Figure 11B), the agglomerates were larger, while in samples with 2.0% (Figure 11C) and 4.0% (Figure 11D) the particles were smaller and more well-defined. This indicates that the presence of larger quantities of silver in the sol-gel oxide modified the material surface, making it more uniform.

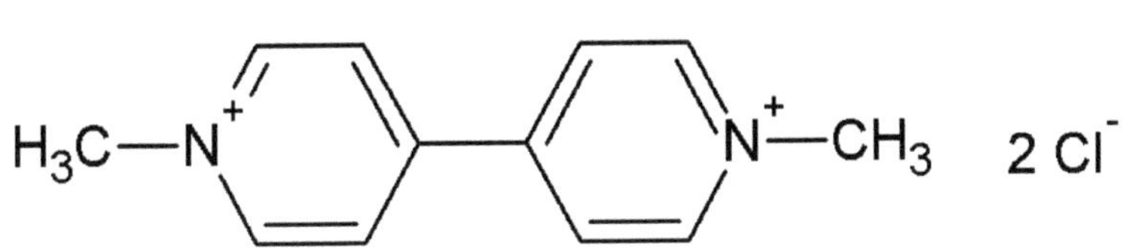

$H_3C—N^+$ ... $N^+—CH_3$ $2\ Cl^-$

Figure 10. Molecular structure of Methyl Viologen.

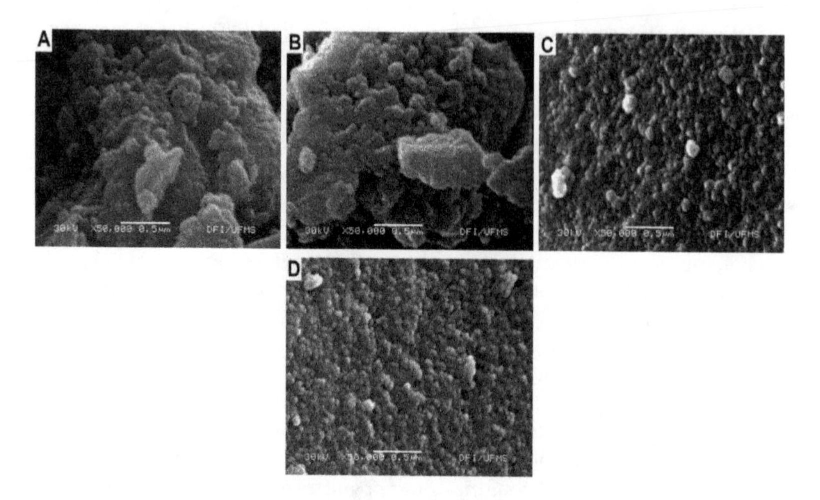

Figure 11. Scanning electron micrographs of: (A) TiO_2 SG 0.5% Ag; (B) TiO_2 SG 1.0% Ag; (C) TiO_2 SG 2.0% Ag; (D) TiO_2 SG 4.0% Ag.

Laser flash photolysis is a technique for producing and investigating excited states and transient reaction intermediates and the kinetics of photochemical reactions. The photocatalytic reduction of MV by TiO_2 or by Ag-doped TiO_2 (2%) in the presence and absence of sodium formate was investigated via the formation of $MV^{\bullet+}$ at different initial concentrations of MV^{2+} (0.05, 0.07, 0.1, 0.15 and 0.2 mmol L^{-1}), monitoring the transient absorption of $MV^{\bullet+}$ at 605 nm [45]. As reported by Tachikawa et al. [45], the transient absorption decays by first order kinetics. The bimolecular electron transfer rate constants in the absence and presence of sodium formate, listed in Table 1, were obtained from linear plots of the observed first-order rate constants versus the concentrations of MV^{2+}. In the presence of sodium formate there is an increase in electron transfer constant for all photocatalysts analyzed; according Tachikawa et al [45], this occurs because the initial oxidation of organic additives, such as sodium formate, generates the $CO_2^{\bullet-}$ radical, which has strong reducing power and can easily reduce other substrates. There is an increase in the electron transfer rate constants in the presence of sodium formate and in the presence of silver, demonstrating the improved efficiency of the oxidation/reduction in the presence of the metal.

	Absence of $NaHCO_2$	Presence of $NaHCO_2$
TiO_2 P25	5.4×10^9 $M^{-1}s^{-1}$	6.0×10^9 $M^{-1}s^{-1}$
TiO_2 P25 2.0% Ag	6.5×10^9 $M^{-1}s^{-1}$	8.0×10^9 $M^{-1}s^{-1}$
TiO_2 SG	3.2×10^9 $M^{-1}s^{-1}$	3.6×10^9 $M^{-1}s^{-1}$
TiO_2 SG 2.0% Ag	4.5×10^9 $M^{-1}s^{-1}$	6.6×10^9 $M^{-1}s^{-1}$

Table 1. Electron transfer rate constants for MV^{2+} in the presence of the different TiO_2 photocatalysts.

To test the photocatalytic activity of the oxides, MV^{2+} photodegradation experiments were performed. The amount of herbicide solution treated was 500 mL and herbicide concentration was determined by spectrophotometric analysis at 250 nm. Titanium dioxide synthesized by the sol-gel method (Figure 12B) had a lower rate of degradation than TiO_2 P25 (Figure 12A). This difference can be attributed to several factors, including the preparation method, crystal structure, surface area, size distribution and porosity. Although the sol-gel oxide had a higher surface area, it contained non-uniform particles of different sizes and therefore had a lower porosity than TiO_2 P25. The oxides synthesized with 2.0% silver showed improved photocatalytic activity for degradation of MV. However, in the presence of oxide doped with 4.0% silver, there was an inhibition of the photocatalytic process, probably due to the excessive amount of silver, which occupied most of the active sites of the catalyst.

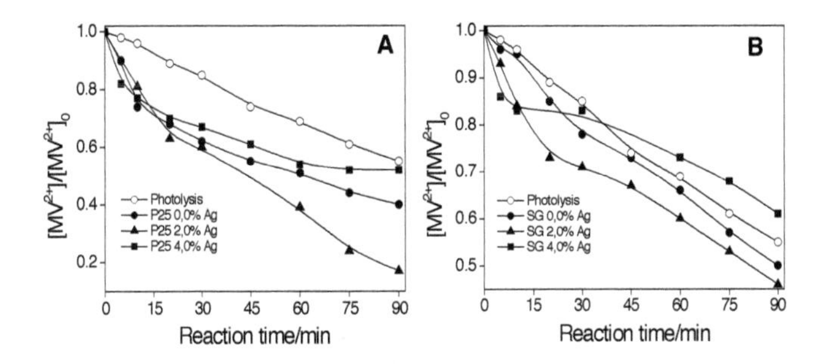

Figure 12. Results of MV^{2+} (0.2 mmol L^{-1}) degradation by heterogeneous photocatalysis (0.5 g of photocatalyst) with (A) TiO_2 P25 and (B) TiO_2 SG.

An alternative method for preparing TiO_2 is via molten-salt synthesis. This approach employs an eutectic mixture of salts, for example NaCl/KCl or $NaNO_3/KNO_3$ in the desired proportion, together with other reagents (oxalates or metals oxides). Molten-salt synthesis [43] was used to prepare TiO_2 using $TiOSO_4.xH_2O.xH_2SO_4$ as the Ti precursor with a melt phase of either $NaNO_3$ or KNO_3. The synthesis reaction occurs according to Equation 14.

$$TiOSO_4 + 2ANO_3 \rightarrow TiO_2[A] + A_2SO_4 + 2NO_2 + \frac{1}{2}O_2 \tag{14}$$

where the symbol [A] indicates the alkali metal cation used in the molten salt ([Na] or [K]).

The oxides synthesized in this manner were characterized by X-ray diffraction and diffuse reflectance spectroscopy. The X-ray diffraction diffractograms (Figure 13) show only the presence of the anatase phase for both oxides synthesized by the molten-salt method. The

E_{gap} values for TiO_2[K] (3.13 eV) and TiO_2[Na] (3.15 eV) were similar to that of P25 (3.13 eV), as expected.

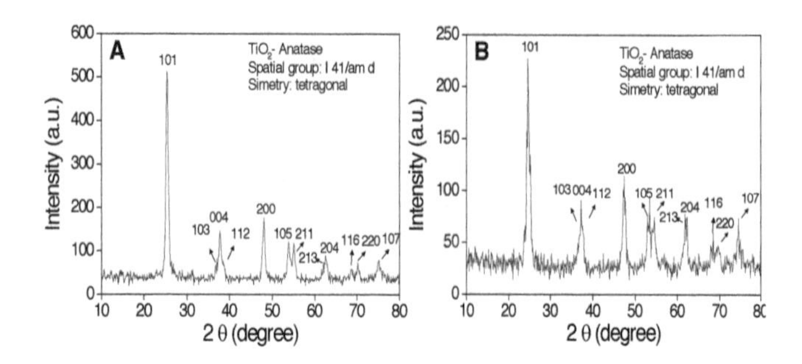

Figure 13. X-ray diffractograms of TiO_2 synthesized by the molten salt method: (A) in $NaNO_3$ and (B) in KNO_3.

The morphologies of the oxides, observed by SEM (Figure 14), exhibited different forms of agglomeration, presumably due to an influence of the alkaline metal nitrate molten salt used in the synthesis.

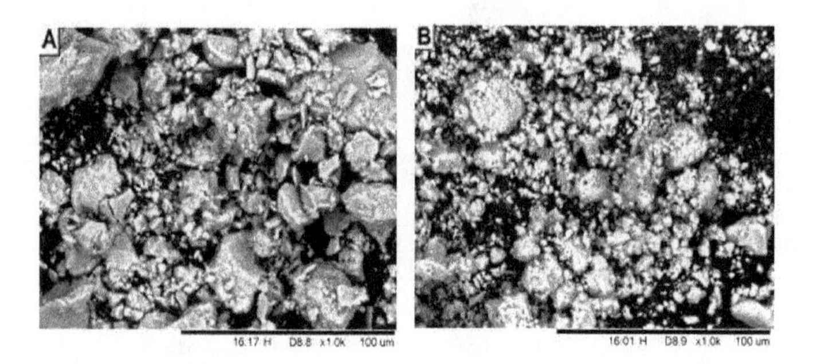

Figure 14. Scanning electron micrographs of (A) TiO_2[K]; (B) TiO_2[Na].

In order to evaluate the photocatalytic activities of the synthesized oxides, they were used for the photodegradation of the dye methylene blue (MB, Figure 15). Although MB is not considered to be a very toxic dye, it can cause harmful effects on living beings. After inhalation, symptoms such as difficulty in breathing, vomiting, diarrhea and nausea may occur in humans [46]. The degradation of MB was carried out in aqueous solution in a 400 mL reactor with an 80 W mercury vapor lamp as the irradiation source. The concentration of MB was determined from its absorption at 654 nm.

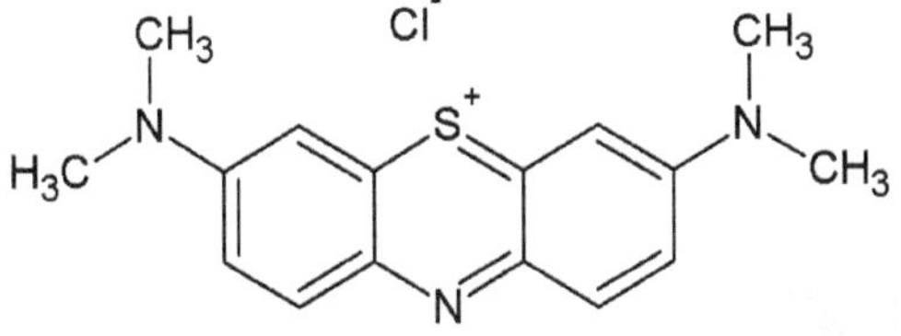

Figure 15. Molecular structure of Methylene Blue.

Figure 16 compares the degradation of MB using the two catalysts synthesized by the molten-salt method. TiO_2[K] produced a net degradation efficiency of 99%. In constrast, TiO_2[Na] degraded only 61% of MB. The lower rate of degradation of MB by TiO_2[Na] and the lower overall efficiency may be related to the differences in aggregation observed in the SEM images of the two oxides (Figure 14). On the basis of these results, TiO_2[K] obtained by the molten salt method would appear to be a promising alternative material for the catalytic photodegradation of organic dyes like MB.

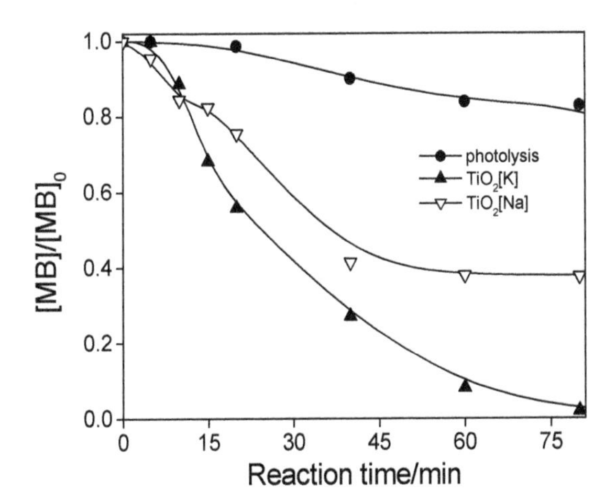

Figure 16. Degradation of MB (0.40 mmol L^{-1}) by heterogeneous photocatalysis using TiO_2 (0.5g L^{-1}) synthesized by the molten salt method.

TiO_2 nanotubes has been subject of several recent studies due to their unique electronic transport properties and their mechanical strength, large surface area and well-defined geometry, which improve their performance in many applications compared to other forms of titanium dioxide. Several studies have reported that highly ordered and uniform TiO_2 nanotubes can be easily obtained using anodization in titanium fluoride [42,47]. The formation of TiO_2 nanotubes by electrochemical anodization is based on a competition between the anodic oxide formation and its dissolution as a soluble complex fluoride.

Ti/TiO_2 electrodes (4 x 2.8 cm) were prepared by anodization of Ti foil using an applied voltage of 20 V in 0.15 mol L^{-1} NH_4F in glycerol (10% H_2O). Self-assembly of nanotubular TiO_2 arrays can be seen on films of Ti obtained under these anodization conditions (Figure 17). The average internal diameter of the nanotubes was 66 nm.

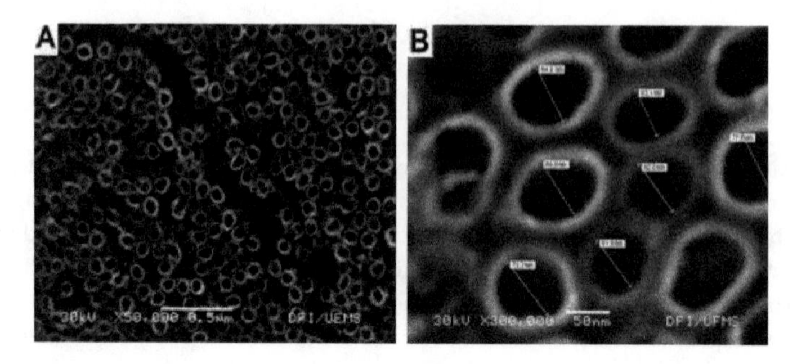

Figure 17. Scanning electron micrographs of TiO_2 nanotubes prepared on a Ti film in 0.15 mol L^{-1} NH_4F in glycerol (10% H_2O) at a potential of 20 V. (A) TiO_2 nanotubes, (B) diameters of the nanotubes.

The photoeletrocatalytic activity of the Ti/TiO_2 electrodes was evaluated by linear voltammetric scans in the potential range of -0.4 to 0.7 V under UV irradiation. The photoanodic current flow arises from the photooxidation of adsorbed water molecules or hydroxyl groups on the titania surface (Figure 18).

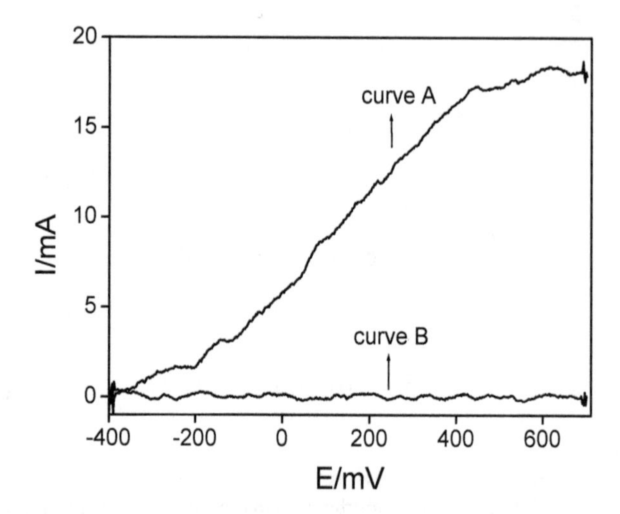

Figure 18. Linear-sweep photovoltammograms for TiO_2 nanotubes on a Ti film in 0.1 mol L^{-1} Na_2SO_4 (curve A) under UV illumination and in the dark (curve B). Scan rate: 5 mV s^{-1}.

Initial studies of photoelectrocatalytic oxidation employing these electrodes was carried out using 2,4 xylidine (Figure 19) as the model pollutant. In 0.1 mol L^{-1} Na_2SO_4 supporting electrolyte applying a potential of 0.6 V, a TOC removal of 62% was obtained.

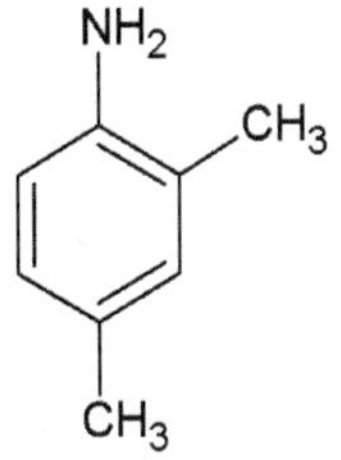

Figure 19. Molecular structure of 2,4-Xylidine.

4. Conclusion

Although there has been a considerable increase in research activity related to advanced oxidation processes (AOP) since 2000, a number of significant challenges must still be overcome to make AOP generally applicable for the treatment of polluted waters and effluents. AOP involving both homogeneous and heterogeneous catalysis have shown good results for degradation of pollutants leading to efficient mineralization. The use of TiO_2 nanoparticles and nanotubes as the photocatalyst have been shown to be viable alternatives for the photodegradation of methylene blue (MB) and for the photoelectrocatalytic oxidation of xylidine. These studies underline the importance of synthesizing new molecules and testing the catalytic efficiencies of novel materials. In addition, new experimental conditions and new AOP technologies need to be developed for the efficient, cost-effective oxidative mineralization of organic materials in polluted waters.

Abbreviations list

AOP	Advanced Oxidation Processes
AOT	Advanced Oxidation Technologies
$HO^{\bullet}$	Hydroxyl Radical
$SO_4^{\bullet-}$	Sulfate Radical Anion
DHB	Dihydroxybenzene
CE	Chlorimurom-Ethyl
MTX	Mitoxantrone
$K_3(FeOx)$	Potassium Ferrioxalate

SEM	Scanning Electron Microscopic
MV^{2+}	Methyl Viologen
MB	Methylene Blue

Acknowledgements

The authors acknowledge the Brazilian funding agencies CAPES, CNPq and FUNDECT for financial and fellowship support. F.H.Q. is associated with NAP-PhotoTech, the USP Research Consortium for Photochemical Technology, and INCT-Catalysis. A.M.Jr. is associated with NAP-PhotoTech and INCT-EMA.

Author details

Amilcar Machulek Jr.[1*], Silvio C. Oliveira[1], Marly E. Osugi[2], Valdir S. Ferreira[1], Frank H. Quina[3], Renato F. Dantas[4], Samuel L. Oliveira[1], Gleison A. Casagrande[5], Fauze J. Anaissi[6], Volnir O. Silva[3], Rodrigo P. Cavalcante[1], Fabio Gozzi[1], Dayana D. Ramos[1], Ana P.P. da Rosa[1], Ana P.F. Santos[1], Douclasse C. de Castro[1] and Jéssica A. Nogueira[1]

*Address all correspondence to: machulekjr@gmail.com

1 Center for Exact Sciences and Technology (CCET), Federal University of Mato Grosso do Sul-UFMS; Campo Grande, MS, Brazil

2 Institute of Chemistry, University of Brasília; Brasília, DF, Brazil

3 Institute of Chemistry and NAP-PhotoTech – USP; University of São Paulo-USP; São Paulo, SP, Brazil

4 Department of Chemical Engineering, Faculty of Chemistry, University of Barcelona; Barcelona, Spain

5 Faculty of Exact Sciences and Technology (FACET), Federal University of Grande Dourados-UFGD; Dourados, MS, Brazil

6 Department of Chemistry, State University of Centro-Oeste - UNICENTRO; Guarapuava, PR, Brazil

References

[1] Legrini, O, Oliveros, E, & Braun, A. M. Photochemical Processes for Water Treatment. Chemical Reviews (1993). , 93(2), 671-698.

[2] Sonntag C vonAdvanced Oxidation Processes: Mechanistic Aspects. Water Science & Technology (2008). , 58(5), 1015-1021.

[3] Matilainen, A, & Sillanpää, M. Removal of Natural Organic Matter from Drinking Water by Advanced Oxidation Processes. Chemosphere (2010). , 80(4), 351-365.

[4] Bauer, R, & Fallmann, H. The Photo-Fenton Oxidation- a Cheap and Efficient Wastewater Treatment Method. Research on Chemical Intermediates (1997). , 23(4), 341-354.

[5] Machulek Jr AQuina F.H., Gozzi F., Silva V.O., Friedrich L.C., Moraes J.E.F. Fundamental Mechanistic Studies of the Photo-Fenton Reaction for the Degradation of Organic Pollutants. In: Puzyn T., Mostrag-Szlichtyng A. (ed) Organic Pollutants Ten Years After the Stockholm Convention- Environmental and Analytical Update. Rijeka: InTech; (2012). , 271-292.

[6] Pignatello, J. J, Oliveros, E, & Mackay, E. Advanced Oxidation Processes for Organic Contaminant Destruction Based on the Fenton Reaction and Related Chemistry [published erratum appears in Critical Reviews in Environmental Science and Technology 2007;37 273-275] Critical Reviews in Environmental Science and Technology (2006). , 36(1), 1-86.

[7] Bossmann, S. H, Oliveros, E, Gob, S, Siegwart, S, & Dahlen, E. P. Payawan Jr L, Straub M, Worner M, Braun AM. New Evidence against Hydroxyl Radicals as Reactive Intermediates in the Thermal and Photochemically Enhanced Fenton Reactions. Journal of Physical Chemistry A (1998). , 102(28), 5542-5550.

[8] Benitez, F. J, Beltran-heredia, J, Acero, J. L, & Rubio, F. J. Chemical Decomposition of Triclorophenol by Ozone, Fenton's Reagent, and UV Radiation. Industrial & Engineering Chemistry Research (1999). , 2(4), 6.

[9] Friedrich, L. C, Mendes, M. A, Silva, V. O, & Zanta, C. Machulek Jr A, Quina FH. Mechanistic Implications of Zinc(II) Ions on the Degradation of Phenol by the Fenton Reaction. Journal of the Brazilian Chemical Society (2012). , 23(7), 1372-1377.

[10] Pontes RFFMoraes JE, Machulek Jr A, Pinto JM. A Mechanistic Kinetic Model for Phenol Degradation by the Fenton Process. Journal of Hazardous Materials (2010).

[11] Zanta CLPSFriedrich LC, Machulek Jr A, Higa KM, Quina FH. Surfactant Degradation by a Catechol-Driven Fenton Reaction. Journal of Hazardous Materials (2010).

[12] Bigda, R. J. Consider Fenton's Chemistry for Wastewater Treatment. Chemical Engineering Progress (1995). , 91(12), 62-66.

[13] Hamilton, G. A, Friedman, J. P, & Campbell, P. M. The Hydroxylation of Anisole by Hydrogen Peroxide in the Presence of Catalytic Amounts of Ferric Ion and Catechol. Scope, Requirements and Kinetic Studies. Journal of the American Chemical Society (1966)., 88(22), 5266-5268.

[14] Hamilton, G. A. Hanifin Jr JW, Friedman JP. The Hydroxylation of Anisole by Hydrogen Peroxide in the Presence of Catalytic Amounts of Ferric Ion and Catechol. Product studies, Mechanism, and Relation to Some Enzymic Reactions. Journal of the American Chemical Society (1966)., 88(22), 5269-5272.

[15] Luna, A. J. Machulek Jr A, Chiavone-Filho O, Moraes JEF, Nascimento CAO. Photo-Fenton Oxidation of Phenol and Organochlorides (DCP and 2,4-D) in Aqueous Alkaline Medium with High Chloride Concentration. Journal of Environmental Management (2012). C) 10-17., 2, 4.

[16] Machulek Jr AMoraes JEF, Okano LT, Silvério CA, Quina FH. Photolysis of Ferric Ion in the Presence of Sulfate or Chloride Ions: Implications for the Photo-Fenton Process. Photochemical & Photobiological Sciences (2009)., 8(7), 985-991.

[17] Machulek Jr AMoraes JE, Vautier-Giongo C, Silverio CA, Friedrich LC, Nascimento CAO, Gonzales MC, Quina FH. Abatement of the Inhibitory Effect of Chloride Anions in the Photo-Fenton Process. Environmental Science & Technology (2007).

[18] Augugliaro, V, Litter, M, Palmisano, L, & Soria, J. The Combination of Heterogeneous Photocatalysis with Chemical and Physical Operations: A Tool for Improving the Photoprocess Performance. Journal of Photochemistry and Photobiology C: Photochemistry Reviews (2006)., 7(4), 127-144.

[19] Machulek Jr AGogritcchiani E, Moraes JE, Quina FH, Oliveros E, Braun AM. Kinetic and Mechanistic Investigation of the Ozonolysis of 2,4-Xylidine (2,4-dimethyl-aniline) in Acid Aqueous Solution. Separation and Purification Technology (2009). , 67(2), 141-148.

[20] Zhao, L, Ma, J, Sun, Z, Liu, Z, & Yang, Y. Experimental Study on Oxidative Decomposition of Nitrobenzene in Aqueous Solution by Honeycomb Ceramic-Catalyzed Ozonation. Frontiers Environmental Science & Engineering in China (2008). , 2(1), 44-50.

[21] Kuns, A, Peralta-zamora, P, Moraes, S. G, & Duran, N. Novas Tendências no Tratamento de Efluentes Têxteis. Química Nova (2002)., 25(1), 78-82.

[22] Lin, S. H, & Yeh, K. L. Looking to Treat Wastewater? Try Ozone. Chemical Engineering (1993)., 100(5), 112-116.

[23] Esplugas, S. Curso técnico, de novembro, (1995). LRR-Universidad de Concepcion-Chile, Concepción, Chile., 8-10.

[24] Pera-titus, M, Garcia-molina, V, Baños, M. A, Gimenez, J, & Esplugas, S. Degradation of Chlorophenols by Means of Advanced Oxidation Processes: a General Review. Applied Catalysis B: Environmental (2004). , 47(4), 219-256.

[25] Straehelin, S, & Hoigné, J. Decomposition of Ozone in Water in the Presence of Organic Solutes Acting as Promoters and Inhibitors of Radical Chain Reactions. Environmental Science & Technology (1985). , 19(12), 1209-1212.

[26] WHO Guidelines for Drinking-Water Quality [electronic resource]: incorporating 1st and 2nd addendaRecommendations. 3rd ed. Geneva, Switzerland, World Health Organization; (2008).

[27] Alfano, O. M, Cabrera, M I, & Cassano, A. E. Photocatalytic Reactions Involving Hydroxyl Radical Attack. Journal of Catalysis (1997). , 172(2), 370-379.

[28] Herrmann, J, Guillard, M, & Pichat, C. P. Heterogeneous Photocatalysis: an Emerging Technology for Water Treatment. Catalysis Today (1993).

[29] Minero, C, Pelizzetti, E, Malato, S, & Blanco, J. Large Solar Plant Photocatalytic Water Decontamination: Effect of Operational Parameters. Solar Energy (1996). , 56(5), 421-428.

[30] Quina, F. H, Nascimento, C. A. O, Teixeira, A. C. S. C, Guardani, R, & Lopez-gejo, J. Degradacion Fotoquimica de Compuestos Organicos de Origen Industrial. In: Nudelman. N. (ed) Quimica Sustentable. Santa Fe, Argentina: Universidad Nacional del Litoral; (2004). , 205-220.

[31] Valente JPSAraujo AB, Bozano DF, Padilha PM, Florentino AO. Síntese e Caracterização Textural do Catalisador CeO_2/TiO_2 Obtido via Sol-Gel: Fotocatálise do Composto Modelo Hidrogenoftalato de Potássio. Eclética Química (2005). , 30(4), 7-12.

[32] Bockelmann, D, Weichgrebe, D, Goslich, R, & Bahnemann, D. Concentrating versus Non-Concentrating Reactors for Solar Water Detoxication. Solar Energy Materials and Solar Cells (1995).

[33] Corminboeuf, C, Carnal, F, Weber, J, Chovelon, J-M, & Chermette, H. Photodegradation of Sulfonylurea Molecules: Analytical and Theoretical DFT Studies. Journal of Physical Chemistry A (2003). , 107(47), 10032-10038.

[34] BrazilMinistério da Agricultura, Pecuária e Abastecimento-MAPA. Instrução Normativa nº 2 de 3 de Janeiro de 2008". Diário Oficial da União de 8 de Janeiro de 2008.;nº 5. Seção 1 (2008). , 5-9.

[35] Turci, R, Sottani, C, Schierl, R, & Minoia, C. Validation Protocol and Analytical Quality in Biological Monitoring of Occupational Exposure to Antineoplastic Drugs. Toxicology Letters (2006).

[36] Cavalcante, R. P, Sandim, L. R, & Bogo, D. Barbosa AMJ, Osugi ME, Blanco M, Oliveira SC, Matos MFC, Machulek Jr A, Ferreira VS. Application of Fenton, Photo-Fenton, Solar Photo-Fenton, and UV/H_2O_2 to Degradation of the Antineoplastic Agent

Mitoxantrone and Toxicological Evaluation. Environmental Science and Pollution Research, in press (DOIs11356-012-1110-y).

[37] Safarzadeh-amiri, A, Bolton, J. R, & Cater, S. R. Ferrioxalate-Mediated Photodegradaation of Organic Pollutants in Contaminated Water. Water Research (1997). , 31(4), 787-798.

[38] Chen, X, & Mao, S. S. Titanium Dioxide Nanomaterials: Synthesis, Properties, Modifications, and Applications. Chemical Reviews (2007). , 107(7), 2891-2959.

[39] Osugi, M. E, Umbuzeiro, G. A, & Anderson, M. A. Zanoni MVB. Degradation of Metallophtalocyanine Dye by Combined Processes of Electrochemistry and Photoelectrochemistry. Electrochimica Acta (2005).

[40] Osugi, M. E, & Umbuzeiro, G. A. Castro FJV, Zanoni MVB. Photoelectrocatalytic Oxidation of Remazol Turquoise Blue and Toxicological Assessment of its Oxidation Products. Journal of Hazardous Materials (2006). , 137(2), 871-877.

[41] Osugi, M. E, & Rajeshwar, K. Ferraz ERA, Oliveira DP, Araújo ÂR, Zanoni MVB. Comparison of Oxidation Efficiency of Disperse Dyes by Chemical and Photoelectrocatalytic Chlorination and Removal of Mutagenic Activity. Electrochimica Acta (2009). , 2009(54), 7-2086.

[42] Osugi, M. E. Zanoni MVB, Chenthamarakshan CR, Tacconi NR, Woldemariam GA, Mandal SS, Rajeshwar K. Toxicity Assessment and Degradation of Disperse Azo Dyes by Photoelectrocatalytic Oxidation on Ti/TiO_2 Nanotubular Array Electrodes. Journal of Advanced Oxidation Technologies (2008). , 11(3), 425-434.

[43] Docters, T, Chovelon, J. M, Hermann, J. M, & Deloume, J. P. Syntheses of TiO_2 Photocatalysts by the Molten Salts Method: Application to the Photocatalytic Degradation of Prosulfuron (R). Applied Catalysis B: Environmental, (2004). , 50(4), 219-226.

[44] Eisler, R. Eisler's Encyclopedia of Environmentally Hazardous Priority Chemicals. Elsevier; (2007).

[45] Tachikawa, T, Tojo, S, Fujitsuka, M, & Majima, T. Direct Observation of the One-Electron Reduction of Metil Viologen Mediated by the CO_2 Radical Anion during TiO_2 Photocatalytic Reactions. Langmuir (2004). , 20(22), 9441-9444.

[46] Mohabansi, N. P, Patil, V. B, & Yenkie, N. A Comparative Study on Photodegradation of Methylene Blue Dye Effluent by Advanced Oxidation Process by Using TiO_2/ZnO Photocatalyst. Rasāyan Journal of Chemistry (2011). , 4(4), 814-819.

[47] Grimes, CA, Mor, GK, & Ti, . $_2$ Nanotube Arrays: Synthesis, Properties, and Applications. Springer; 2009.

Chapter 9

Photocatalytic Degradation of Organic Pollutants in Water

Muhammad Umar and Hamidi Abdul Aziz

Additional information is available at the end of the chapter

1. Introduction

A photocatalyst is defined as a substance which is activated by adsorbing a photon and is capable of accelerating a reaction without being consumed [1]. These substances are invariably semiconductors. Semiconducting oxide photocatalysts have been increasingly focused in recent years due to their potential applications in solar energy conversion and environmental purification. Semiconductor heterogeneous photocatalysis has enormous potential to treat organic contaminants in water and air. This process is known as advanced oxidation process (AOP) and is suitable for the oxidation of a wide range of organic compounds. Among AOPs, heterogeneous photocatalysis have been proven to be of interest due to its efficiency in degrading recalcitrant organic compounds. Developed in the 1970s, heterogeneous photocatalytic oxidation has been given considerable attention and in the past two decades numerous studies have been carried out on the application of heterogeneous photocatalytic oxidation process with a view to decompose and mineralize recalcitrant organic compounds. It involves the acceleration of photoreaction in the presence of a semiconductor catalyst [2]. Several semiconductors (TiO_2, ZnO, Fe_2O_3, CdS, ZnS) can act as photocatalysts but TiO_2 has been most commonly studied due to its ability to break down organic pollutants and even achieve complete mineralization. Photocatalytic and hydrophilic properties of TiO_2 makes it close to an ideal catalyst due to its high reactivity, reduced toxicity, chemical stability and lower costs [3]. Fujishima and Honda [4] pioneered the concept of titania photocatalysis (also known as "Honda-Fujishima effect"). Their work showed the possibility of water splitting in a photoelectrochemical cell containing an inert cathode and rutile titania anode. The applications of titania photoelectrolysis has since been greatly focused in environmental applications including water and wastewater treatment. This chapter provides insight into the fundamentals of the TiO_2 photocatalysis, discusses the effect of variables affecting the performance of degradation of organic pollutants in water with a view to current state of knowledge and future needs.

2. Mechanism and fundamentals of photocatalytic reactions

Heterogeneous photocatalysis using UV/TiO_2 is one of the most common photocatalytic process and is based on adsorption of photons with energy higher than 3.2 eV (wavelengths lower than ~390 nm) resulting in initiating excitation related to charge separation event (gap band) [5]. Generation of excited high-energy states of electron and hole pairs occurs when wide bandgap semiconductors are irradiated higher than their bandgap energy. It results in the promotion of an electron in the conductive band (e_{CB}^-) and formation of a positive hole in the valence band (h_{VB}^+) [5] (Eq. 1). The h_{VB}^+ and e_{CB}^- are powerful oxidizing and reducing agents, respectively. The h_{VB}^+ reacts with organic compounds resulting in their oxidation producing CO_2 and H_2O as end products (Eq. 2). The h_{VB}^+ can also oxidize organic compounds by reacting with water to generate $^{\bullet}OH$ (Eq. 3). Hydroxyl radical ($^{\bullet}OH$) produced by has the second highest oxidation potential (2.80 V), which is only slightly lower than the strongest oxidant – fluorine. Due to its electrophilic nature (electron preferring), the $^{\bullet}OH$ can non-selectively oxidize almost all electron rich organic molecules, eventually converting them to CO_2 and water (Eq. 4).

$$TiO_2 + hv\ (<387\ nm) \rightarrow e_{CB}^- + h_{VB}^+ \tag{1}$$

$$h_{VB}^+ + R \rightarrow \text{intermediates} \rightarrow CO_2 + H_2O \tag{2}$$

$$H_2O + h_{VB}^+ \rightarrow {}^{\bullet}OH + H^+ \tag{3}$$

$$^{\bullet}OH + R \rightarrow \text{intermediates} \rightarrow CO_2 + H_2O \tag{4}$$

where R represents the organic compound.

The conductive band can react with O_2 forming an anion radical superoxide as shown in Eq. 5. Further reaction can lead to the formation of hydrogen peroxide which lead to the formation of $^{\bullet}OH$ [6]. The mechanism of the electron hole-pair formation when the TiO_2 is irradiatied is given in Figure 1 [7].

$$e_{CB}^- + O_2 \rightarrow O_2^{\bullet -} \tag{5}$$

The presence of dissolved oxygen is extremely important during photocatalytic degradation as it can make the recombination process on TiO_2 (e_{CB}^-/h_{VB}^+) difficult which results in maintaining the electroneutrality of the TiO_2 particles [5]. In other words, it is important for effective photocatalytic degradation of organic pollutants that the reduction process of oxygen and the oxidation of pollutants proceed simultaneously to avoid the accumulation of electron in the conduction band and thus reduce the rate of recombination of e_{CB}^- and h_{VB}^+ [8, 9].

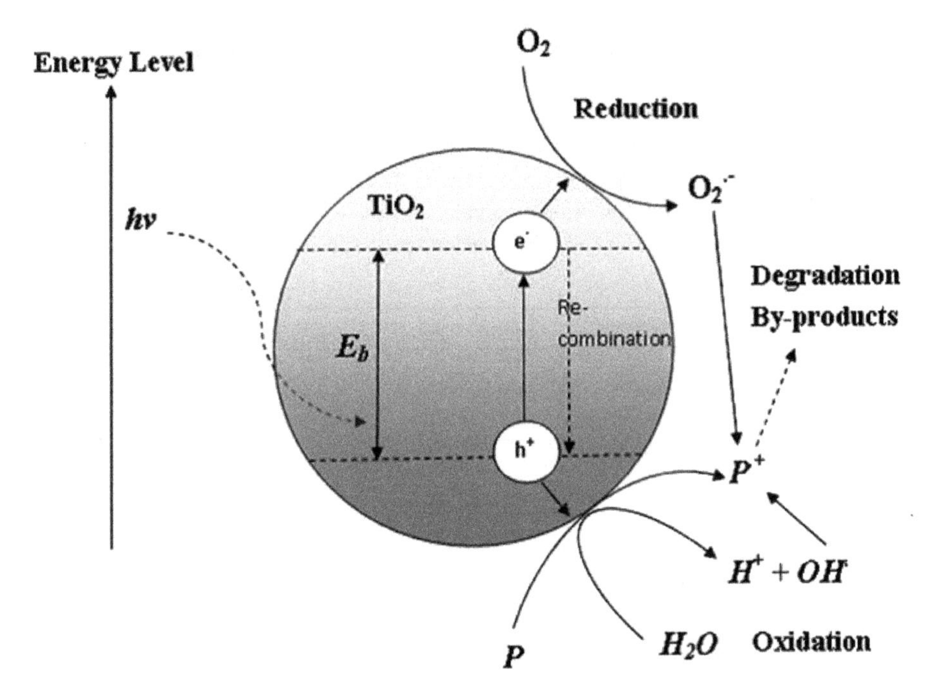

Figure 1. Mechanism of electron-hole pair formation in a TiO_2 particle in the presence of pollutant in water [7].

3. Types of photocatalysts and their characteristics

A number of solids can be referred to as photocatalysts and as mentioned earlier, metal oxide semiconductors are considered to be the most suitable photocatalysts due to their photocorrosion resistance and wide band gap energies [10]. Table 1 provides the band gap energies at corresponding wavelength for well known semiconductors. TiO_2 stands out as the most effective photocatalyst and has been extensively used in water and wastewater treatment studies because it is cost effective, thermally stable, non-toxic, chemically and biologically inert and is capable of promoting oxidation of organic compounds [11]. The photocatalytic activity of TiO_2 is dependent on surface and structural properties which include crystal composition, surface area, particle size distribution, porosity and band gap energy [12]. TiO_2 is also known as titania, titanic oxide, titanium white, titanic anhydride, or titanic acid anhydride. It is prepared using ilmenite and rutile in crystalline forms called anatase and rutile. The anatase form is achieved by processing of titanium sulphate, which is achieved when ilmenite is treated with sulphuric acid. Rutile crystalline form is obtained when raw rutile is chlorinated and the resulting titanium tetrachloride is subjected to vapor phase oxidation [13]. When photon energy (hv) of higher than or equal to the bandgap energy of TiO_2 is illuminated onto its surface, typically 3.2 eV (anatase) or 3.0 eV (rutile), the lone electron is photoexcited to the empty conduction band in femtoseconds [7]. Degussa P25 which is the most widely used form of TiO_2 is composed of 75% anatase and 25% rutile and has a specific BET surface area of 50 m^2/g. The high effectiveness of D25 is related to the inhibition of recombination process on TiO_2 (e_{CB}^-/h_{VB}^+) due to the smaller band gap of rutile that absorbs photons and generates electron-hole pairs and the electron transfer from the rutile conductive band to the electron traps occurs in the anatase phase [14].

Semiconductor	Band gas energy (eV)	Wavelength
TiO_2 (rutile)	3.0	413
TiO_2(anatase)	3.2	388
ZnO	3.2	388
ZnS	3.6	335
CdS	2.4	516
Fe_2O_3	2.3	539

Table 1. Band gap energies of various semiconductors at relevant wavelengths [15]

4. Radiation sources for photocatalysis

Both artificial UV lamps and sunlight can be used as the radiation source for photocatalytic process. Artificial UV lamps containing mercury are the most commonly used source of UV irradiation. These can be divided into low pressure mercury lamp, medium pressure mercury lamp and high pressure mercury lamp. Sunlight has also been used in the photocatalytic process as nearly 4-5% of the sunlight that reaches the earth's surface is in the 300-400 nm near UV light range. Furthermore solar energy has limitations due to the graphical variations when compared with the artificial UV lamps. However ongoing interests and developments in harnessing solar energy are expected to increase its use in photocatalytic degradation applications.

5. Photocatalytic reactors

Photocatalytic reactors can be classified based on the deployed state of the photocatalyst, i.e., suspended or attached. Photocatalytic reactors can use either UV or solar radiation. Solar photocatalytic reactors have been of great interest for the photoxodation of organic contaminants in water. Such kind of reactors can be divided into concentrating or non-concentrating reactors [16]. Both the reactor types extend certain advantages and disadvantages. For example, non-concentrating reactors have negligible optical losses and therefore can use direct and diffuse sun irradiation but are larger in size compared with the concentrating reactors and have high frictional pressure losses [16]. However, the use of solar radiated photoreactors is limited due to the intrinsic nature of the TiO_2 particles. Following section provides details on the type of reactors used in various studies for the degradation of organic pollutants in water.

5.1. Slurry reactors

Until recently, TiO_2 slurry reactors are most commonly type used in water treatment. These show largest photocatalytic activity compared with the immobilized photocatalyst and provide a high total surface area of photocatalyst per unit volume which is one of the most important factor configuring a photocatalytic reactor [7]. However, these reactors require separation of the sub-micron TiO_2 particles from the treated water which complicates the treatment process. Several techniques were proposed to achieve post-treatment separation such as the use of settling tanks (overnight particle settling) or external cross-flow filtration system [7]. However the use of filtrations systems increases the cost of the treatment process.

5.2. Immobilized TiO_2 reactors

Photocatalytic reactors with immobilized TiO_2 are those in which catalyst is fixed to support via physical surface forces or chemical bonds. These reactors extend the benefit of not requiring catalyst recovery and permit the continuous use of the photocatalyst [16]. Hybrid photocatalytic membrane reactors have been developed to achieve the purpose of downstream separation of photocatalyst. The photocatalytic membrane reactors can be generalized in two categories (1) irradiation of the membrane module and (2) irradiation of feed tank containing photocatalyst in suspension [17]. Various membranes such as microfiltration, ultrafiltration, and nanofiltration membranes may be used for this purpose depending on the requirements of the treated water quality [7]. Photocatalytic membrane reactors have been successfully used for the degradation of tricholoroethylene and 4-nitrophenol [18, 19]. However, these reactors possess drawbacks such as low surface area to volume ratios, catalyst fouling and significant pressure drop [16]. Another problem associated with the membrane photocatalytic reactors is the diffusion of organic compounds to the catalyst surface which is slow particularly when the organic compounds concentration is low [20]. One possible solution to the slow diffusion is using pores of nano size to enable photocatalyst to perform selective permeation and to produce an oxidized permeate stream [21].

It can be observed that the photocatalytic reactors can be either slurry or immobilized systems and each possess certain advantages and disadvantages related to their design and efficiency. Further research on the design and energy efficiency of photocatalytic reactors could make photocatalytic degradation process more feasible for future applications in water treatment. Membrane photoreactors appear to be a promising alternative to conventional photoreactors and more research in this area can assist overcome some of the problems faced with the use of conventional reactors.

6. Factors affecting the degradation performance

6.1. Catalyst loading

The amount of TiO_2 being directly proportional to the overall photocatalytic reaction rate, the concentration of the TiO_2 particles affects the overall photocatalysis reaction rate in a true heterogeneous catalytic regime [2]. However, when the amount of TiO_2 is above certain level (saturation stage), the light photon adsorption co-efficient decreases radially and the excess photocatalyst can create a light screening effect that leads to the reduction in the surface area exposed to irradiation and thus reduces the photocatalytic efficiency of the process [7]. A number of studies have reported the effect of TiO_2 loadings on the treatment efficiency of the photocatalytic reactor [2, 22-24]. Although a direct comparison between these studies is difficult to be made due to the differences in the working geometry, radiation fluxes and wavelengths used, it was evident that the optimum dosages of photocatalyst loading were dependent on the dimension of the reactor. The importance of the determination of the reactor diameter has been emphasized to achieve effective photon absorption [25]. The optimum dosage of TiO_2 used by various authors either alone or in combination with other catalysts is given in Table 2.

6.2. pH of the solution

The effect of pH on the photocatalytic reaction has been extensively studied [26, 27] due to the fact that photocatalytic water treatment is highly dependent on the pH as it affects the charge on the catalyst particles, size of aggregates and the position of conductance and valance bands [7]. Furthermore the surface of the TiO_2 can be protonated or deprotonated under acidic or alkaline conditions [2], respectively according to the reaction given below.

$$TiOH + H^+ \rightarrow TiOH_2^+ \tag{6}$$

$$TiOH + OH^- \rightarrow TiO^- + H_2O \tag{7}$$

The point of zero discharge for P25 Degussa, the most commonly used form of TiO_2 is 6.9 [28]. Therefore the surface of the TiO_2 is positively charged under acidic conditions and negatively charged under alkaline conditions. The maximum oxidizing capacity of the titania is at lower pH however the reaction rate is known to decrease at low pH due to excess H^+ [29]. The selection of pH is thus need to be appropriate in order to achieve maximum degradation efficiency.

Target compound	Photocatalyst	Optimum dosage (g/L)	References
Erioglaucine	TiO_2	0.3	[23]
Tebuthioron	TiO_2	5	[33]
Propham	TiO_2	5	[33]
Triclopyr	TiO_2	2	[34]
Phorate	TiO_2	0.5	[54]
Turbophos	TiO_2	0.5	[55]
Trichlorfon	TiO_2	8	[56]
Methamodiphos	Re- TiO_2	1	[57]
Methylene blue	La-Y/TiO_2	4	[58]
Carbendazim	TiO_2	0.07	[59]
Direct red 23	Ag-TiO_2	3	[60]
Phenol	Pr- TiO_2	1	[61]
Carbofuran	TiO_2	0.1	[62]
Beta-cypermethrin	RuO_2-TiO_2	5	[63]
Aniline	Pt-TiO_2	2.5	[64]
Benzylamine	Pt-TiO_2	2.5	[64]
Glyphosate	TiO_2	6	[65]
Picloram	TiO_2	2	[66]
Floumeturon	TiO_2	3	[67]
Imazapyr	TiO_2	2.5	[68]

Table 2. Optimum dosage of photocatalyst for degradation of organic compounds

6.3. Size and structure of the photocatalyst

Surface morphology such as particle size and agglomerate size, is an important factor to be considered in photocatalytic degradation process because there is a direct relationship between organic compounds and surface coverage of the photocatalyst [30]. The number of photon striking the photocatalyst controls the rate of reaction which signifies that the reaction takes place only in the absorbed phase of the photocatalyst [2, 31]. A number of different forms of TiO_2 have been synthesized to achieve the desired characteristics of the photocatalyst [32]. Some of the examples include UV100, PC500 and TTP. For the degradation of various organic compound such as pesticides and dyes, the efficacy of these photocatalysts has generally been reported in the order of Degussa P25 > UV100 > PC500 >TTP [33-36].

6.4. Reaction temperature

An increase in reaction temperature generally results in increased photocatalytic activity however reaction temperature >80°C promotes the recombination of charge carriers and disfavor the adsorption of organic compounds on the titania surface [2]. A reaction temperature below 80°C favours the adsorption whereas further reduction of reaction temperature to 0°C results in an increase in the apparent activation energy [7]. Therefore temperature range between 20-80°C has been regard as the desired temperature for effective photomineralization of organic content.

6.5. Concentration and nature of pollutants

The rate of photocatalytic degradation of certain pollutant depends on its nature, concentration and other existing compounds in water matrix. A number of studies have reported the dependency of the TiO_2 reaction rate on the concentration of contaminants in water [37]. High concentration of pollutants in water saturates the TiO_2 surface and hence reduces the photonic efficiency and deactivation of the photocatalyst [38]. In addition to the concentration of pollutants, the chemical structure of the target compound also influences the degradation performance of the photocatalytic reactor. For example, 4-chlorophenol requires prolonged irradiation time due to its transformation to intermediates compared with oxalic acid that transforms directly to carbon dioxide and water, i.e., complete mineralization [39]. Furthermore if the nature of the target water contaminants is such that they adhere effectively to the photocatalyst surface the process would be more effective in removing such compounds from the solution. Therefore the photocatalytic degradation of aromatics is highly dependent on the substituent group [2]. The organic substrates with electron withdrawing nature (benzoic acid, nitrobenzene) strongly adhere to the photocatalyst and therefore are more susceptible to direct oxidation compared with the electron donating groups [40].

6.6. Inorganic ions

Various inorganic ions such as magnesium, iron, zinc, copper, bicarbonate, phosphate, nitrate, sulfate and chloride present in wastewater can affect the photocatalytic degradation rate of the organic pollutants because they can be adsorbed onto the surface of TiO_2 [41-43]. Photocatalytic deactivation has been reported whether photocatalyst is used in slurry or fixed-bed configuration which is related to the strong inhibition from the inorganic ions on the surface of the TiO_2 [44]. A number of studies have been conducted on the effect of inorganic ions (anions and cations) on TiO_2 photocatalytic degradation [30, 45-51]. Some of the cations such as copper, iorn and phosphate have been reported to decrease the photodegradation efficiency if they are present at certain concentrations whereas calcium, magnesium and zinc have little effect on the photodegradation of organic compounds which is associated to the fact that these cations have are at their maximum oxidation states that results in their inability to have any inhibitory effect on the degradation process [7].

The inorganic anions such as nitrate, chlorides, carbonates and sulphates are also known to inhibit the surface activity of the photocatalyst. The presence of salts diminishes the colloidal stability, increases mass transfer and reduces the surface contact between the pollutant and the photocatalyst [7]. Other than fouling of the TiO_2 surface certain anions such as chlorides, carbonates, phosphate and sulphates also scavenge both the hole and the hydroxyl radicals [52]. The mechanism of hole and radical scavenging by chloride has been proposed by Matthews and McEnvoy [53] as follows.

$$Cl^- + {}^{\bullet}OH \rightarrow Cl^{\bullet} + OH^- \quad (8)$$

$$Cl^- + h^+ \rightarrow Cl^{\bullet} \quad (9)$$

The inhibitory effect of chloride ions occurs through preferential adsorption displacement mechanism which results in reducing the number of OH^- available on the photocatalyst surface [7].

The fouling of photocatalytic surface can be reduced by pre-treatment of water such as with ion exchange resins which have been reported to reduce the fouling and so the cost of treatment (Burns et al., 1999). Similarly the fouling induced by sulphates and phosphates has been reported to be displaced by NaOH, KOH and $NaHCO_3$ [41]. However, most of studies conducted on the effect of inorganic ions are based on the model compounds and therefore do not necessarily represent their effect in real water matrix where several ions exist. More work concentrating on the effect of complex mixtures of inorganic ions is thus required.

7. Conclusions

Photocatalytic degradation of organic pollutants is promising technology due to its advantage of degradation on pollutants instead of their transformation under ambient conditions. The process is capable of removing a wide range of organic pollutants such as pesticides, herbicides, and micropollutants such as endocrine disrupting compounds. Although significant amount of research has been conducted on TiO_2 photocatalysis at laboratory scale, its application on industrial scale requires certain limitations to be addressed. However the application of this treatment is constrained by several factors such as wide band gap (3.2eV), lack and inability of efficient and cost-effective catalyst for high photon-efficiency to utilize wider solar spectra. The effect of variables is required to be further studied in real water matrix to achieve representative results. The results achieved can be used to optimize the process and design appropriate reactor for potential large scale applications. The use of solar radiation has to be improved by virtue of the design of the photoreactor in order to reduce the cost of treatment. Further research to investigate the degradation of the real water constituents is required to better comprehend the process applications.

Author details

Muhammad Umar[1] and Hamidi Abdul Aziz[2*]

*Address all correspondence to: cehamidi@usm.eng.my

1 School of Civil, Environmental and Chemical Engineering, RMIT University, Melbourne, Victoria, Australia

2 School of Civil Engineering, University Sains Malaysia, Engineering Campus, Nibong Tebal, Penang, Malaysia

References

[1] Fox M., Photocatalytic Oxidation of Organic Substances. In: Kluwer (ed.) Photocatalysis and Environment: Trends and Applications. New York Academic Publishers: 1988. p. 445–467.

[2] Gaya U.I., Abdullah A.H. Heterogeneous Photocatalytic Degradation of Organic Contaminants over Titanium Dioxide: A Review of Fundamentals, Progress and Problems. Journal of Photochemistry and Photobiology C Photochemistry Reviews 2008 (9) 1-12.

[3] Fujishima A., Rao T.N., Tryk D.A. Titanium Dioxide Photocatalysis. Journal of Photochemistry and Photobiology C Photochemistry Reviews 2000 (1) 1-21.

[4] Fujishima A., Honda K. Electrochemical Photolysis of Water at a Semiconductor Electrode. Nature 1972 (238) 37-38.

[5] Boroski M., Rodrigues A.C., Garcia J.C., Sampaio L.S., Nozaki J., Hioka N. Combined Electrocoagulation and TiO_2 Photoassisted Treatment Applied to Wastewater Effluents from Pharmaceutical and Cosmetic Industries. Journal of Hazardous Materials 2009 (162) 448–454.

[6] Pirkanniemi K., Sillanpaa M. Heterogeneous Water Phase Catalysis as an Environmental Application: A Review. Chemosphere 2000 (48) 1047–1060.

[7] Chong M.N., Jin B., Chow C.W.K., Saint C. Recent Developments in Photocatalytic Water Treatment Technology: A Review. Water Resources 2010 (44) 2997-3027.

[8] Hoffmann M.R., Martin S.T., Choi W., Bahnemann D.W. Environmental Applications of Semiconductor Photocatalysis. Chemical Reviews 1995 (95) 69-96.

[9] Herrmann J.M. Heterogeneous Photocatalysis: Fundamentals and Applications to the Removal of Various Types of Aqueous Pollutants. Catalysis Today 53 (1999) 115-129.

[10] Fox M.A., Dulay M.T. Heterogeneous Photocatalysis. Chemical Reviews 1993 (93) 341–357.

[11] Mandelbaum P., Regazzoni A., Belsa M., Bilme S. Photo-electron-oxidation of Alcohol on Titanium Dioxide Thin Film Electrodes. Journal of Physics and Chemistry B. 1999 (103) 5505-5511.

[12] Ahmed S., Rasul M.G., Brown R., Hashib M.A. Influence of Parameters on the Heterogeneous Photocatalytic Degradation of Pesticides and Phenolic Contaminants in Wastewater: A Short Review. Journal of Environmental Management 2011 (92) 311-330.

[13] Hawley G. The Condensed Chemical Dictionary. 8th Ed. (revised). Litton Educational Publishing Incorporation, 1971.

[14] Hurun D.C., Agrios A.G., Gray K.A., Rajh T., Thurnaur M.C. Explaining the Enhanced Photocatalytic Activity of Degussa P 25 Mixed-phase TiO_2 using EPR. The Journal Physics and Chemistry B 2003 (107) 4545-4549.

[15] Rajeshwar, K., Ibanez, J. Environmental Electrochemistry, Fundamentals and Fundaments in Pollution Abatement. Acadmic Press, San Diego 1997.

[16] de Lasa H., Serrano B., Salaices M. Photocatalytic Reaction Engineering. Springer Science: USA 2005.

[17] Molinari R., Palmisano L., Drioli E., Schiavello M. Studies on Various Reactor Configurations for Coupling Photocatalysis and Membrane Processes in Water Purification. Journal of Membrane Science 2002 (206) 399–415.

[18] Artale M.A., Augugliaro V., Drioli E., Golemme G., Grande C., Loddo V., Molinari R., Palmisano L., Schiavello M. Preparation and Characterisation of Membranes with Entrapped TiO_2 and Preliminary Photocatalytic Tests. Annali di Chimica 2000 (91) 127–136.

[19] Tsuru T.,Toyosada T., Yoshioka T., Asaeda M. Photocatalytic Membrane Reactor using Porous Titanium Dioxide Membranes. Journal of Chemical Engineering Japan 2003 (36) 1063–1069.

[20] Augugliaro V., Litter M., Palmisano L., Soria J. The Combination of Heterogeneous Photocatalysis with Chemical and Physical Operations: A tool for Improving the Photoprocess Performance. Journal of Photochemistry and Photobiology C: Photochemistry Reviews 2006 (7) 127–144

[21] Herz R.K. Intrinsic Kinetics of First-order Reactions in Photocatalytic Membranes and Layers. Chemical Engineering Japan 2004 (99) 237–245.

[22] Chin S.S., Chiang K., Fane A.G. The Stability of Polymeric Membranes in TiO_2 Photocatalysis process. Journal of Membrane Science 2006 (275) 202-211.

[23] Daneshvar N., Salari D., Niaei A., Khataee A. R. Photocatalytic Degradation of the Herbicide Erioglaucine in the Presence of nanosized titanium dioxide: Comparison

and Modeling of Reaction Kinetics. Journal of Environmental Science and Health Part B: Pesticides, Food Contaminants and Agricultural Wastes. 2006 (41) 1273-1290.

[24] Chong M.N., Jin B., Zhu H.Y., Chow C.W.K., Saint C. Application of H-titanate Nano Fibers for Degradation of Congo Red in an Annular Slurry Photoreactor. Chemical Engineering Journal 2009 (150) 49-54.

[25] Malato S., Fernández-Ibánez P., Maldonado M.I., Blanco J., Gernjak W. Decontamination and Disinfection of Water by Solar Photocatalysis: Recent Overview and Trends. Catalysis Today 2009 (147) 1-59.

[26] Mrowetz M., Selli E. Photocatalytic Degradation of Formic and Benzoic Acids and Hydrogen Peroxide Evolution in TiO_2 and ZnO Water Suspensions. Journal of Photochemistry and Photobiology A: Chemistry 2006 (180) 15-22.

[27] Wang W-Y., Ku Y. Effect of solution pH on the adsorption and photocatalytic reaction behaviors of dyes using TiO_2 and Nafion-coated TiO_2. Colloids and Surfaces A Physicochemical and Engineering Aspects. 2007 (302) 261-268.

[28] Kosmulski M. pH-dependent Surface Charging and Points of Zero Charge: III. Update. Journal of Colloid Interface Science 2006 (298) 730-741.

[29] Sun J., Wang S., Sun J., Sun R., Sun S., Qiao, L. Photocatalytic Degradation and Kinetics of Orange G using Nano-sized Sn(IV)/TiO_2/AC Photocatalyst. Journal of Molecular Catalysis A: Chemical 2006 (260) 241-246.

[30] Guillard C., Lachheb H., Houas A., Elaloui E., Hermann J-M. Influence of Chemical Structure of Dyes, of pH and of Inorganic Salts on their Photocatalytic Degradation by TiO_2 Comparison of the Efficiency of Powder and Supported TiO_2. Journal of Photochemistry and Photobiology A: Chemical 2003 (158) 27-36.

[31] Kogo K., Yoneyama H., Tamura H. Photocatalytic oxidation of cyanide on platinized titanium dioxide. The *Journal of Physical Chemistry1980* (84) 1705–1710.

[32] Gao Y., Liu H. Preparation and Catalytic Property Study of a Novel kind of Suspended Photocatalyst of TiO_2-activated Carbon Immobilized on Silicone Rubber Film. Materials Chemistry and Physics 2005 (92) 604-608.

[33] Muneer M., Qamar M., Saquib M., Bahnemann D. Heterogeneous Photocatalysed Reaction of three Selected Pesticide Derivatives, Propham, Propachlorand Tebuthiuron in Aqueous Suspensions of Titanium Dioxide. Chemosphere 2005 (61) 457-468.

[34] Qamar M., Muneer M., Bahnemann D. Heterogeneous Photocatalysed Degradation of two Selected Pesticide Derivatives, Triclopyr and Daminozid Inaqueous Suspensions of Titanium Dioxide. Journal of Environmental Management 2006 (80) 99-106.

[35] Tariq M.A., Faisal M. Muneer M. Semiconductor-mediated Photocatalysed Degradation of two selected Azo Dye Derivatives, Amaranth and Bismarck Brown in Aqueous Suspension. Journal of Hazardous Materials B 2005 (127) 172–179

[36] Tariq M.A., Faisal M., Saquib M., Muneer M. Heterogeneous photocatalytic degradation of ananthraquinone and a triphenylmethane dye derivative in aqueous suspensions of semiconductor. Dyes and Pigments 2008 (76) 358-365.

[37] Chong M.N., Lei S., Jin B., Saint C., Chow C.W.K. Optimisation of an Annular Photoreactor Process for Degradation of Congo Red using a newly Synthesized Titaniaim Pregnated Kaolinite Nano-photocatalyst. Separation and Purification Technology 2009 (67) 355-363.

[38] Saquib M., Muneer M. TiO_2-mediated Photocatalytic Degradation of a Triphenyl Methane Dye (Gentian Violet), in Aqueous Suspensions. Dyes and Pigments 2003 (56) 37-49.

[39] Bahnemann D. Photocatalytic Water Treatment: Solar Energy Applications. Solar Energy 2004 (77) 445-459.

[40] Bhatkhnade D.S., Kamble S.P., Sawant S.B., Pangarkar V.G. Photocatalytic and photochemical degradation ofnitrobenzene using artificial ultraviolet light. Chemical Engineering Journal 2004 (102) 283-290.

[41] Abdullah M., Low G., Mathews R.W. Effects of common inorganic ions on rates of photocatalytic oxidation of organic carbon over illuminated titanium dioxide. Journal of Physical Chemistry 1990 (94) 6820.

[42] Lin C., Lin K. Photocatalytic Oxidation of Toxic Organohalides withTiO_2/UV: The effects of Humic Substances and Organic Mixtures. Chemosphere 2007 (66) 1872–1877.

[43] Parent Y., Blake D., Magrini-Bair K., Lyons C., Turchi C., Watt A., Wolfrum E., Praire M. 1996. Solar Photocatalytic Process for the Purification of Water: State of Development and Barriers to Commercialization. Solar Energy 1996 (56) 429–437.

[44] Crittenden J.C., Zhang Y., Hand D.W., Perram D.L., Marchand E.G. Solar detoxification of fuel-contaminated groundwater using fixed-bed photocatalysts. Water Environment Research 1996 (68) 270-278.

[45] Habibi M.H., Hassanzadeh A., Mahdavi S. The effect of operational parameters on the photocatalytic degradation of three textile azo dyes in aqueous TiO_2 suspensions. Journal of Photochemistry and Photobiology A: Chemical 172 (2005) 89-96.

[46] Leng W., Liu H., Cheng S., Zhang J., Cao C. Kinetics ofphotocatalytic degradation of aniline in water over TiO_2 supported on porous nickel. Journal of Photochemistry and Photobiology A: Chem. 2000 (131) 125-132.

[47] Özkan A., Özkan M.H., Gürkan R., Akçay M., Sökmen M. Photocatalytic Degradation of a Textile Azo Dye, Sirius Gelb GC on TiO_2 or AgeTiO_2 Particles in the Absence and Presence of UV Irradiation: The Effects of Some Inorganic Anions on the Photocatalysis. Journal of Photochemistry and Photobiology A: Chemical 2004 (163) 29-35.

[48] Riga A., Soutsas K., Ntampegliotis K., Karayannis V., Papapolymerou G. Effect of System Parameters and of Inorganic Salts on the Decolorization and Degradation of

Procion H-exldyes. Comparison of H_2O_2/UV, Fenton, UV/ Fenton, TiO_2/UV and TiO_2/UV/H_2O_2 Processes. Desalination 2007 (211) 72-86.

[49] Rincón A.G., Pulgarin C. Effect of pH, Inorganic Ions, Organic Matter and H_2O_2 on E. coli K12 Photocatalytic Inactivation by TiO_2-implications in Solar Water Disinfection. Applied Catalysis B: Environmental 2004 (51) 283-302.

[50] Schmelling D.C., Gray K.A., Vamat P.V. The Influence of Solution Matrix on the Photocatalytic Degradation of TNT in TiO_2 slurries. Water Resources 1997 (31) 1439-1447

[51] Wang K., Zhang J., Lou L., Yang S., Chen Y. UV or Visible Light Induced Photodegradation of AO7 on TiO_2 Particles: The Influence of Inorganic Anions. Journal of Photochemistry Photobiology A: Chemical 165 (2004) 201-207.

[52] Diebold U. The Surface Science of Titanium Dioxide. Surface Science Reports 2003 (48) 53-229.

[53] Matthews R.W., McEnvoy S.R. Photocatalytic Degradation of Phenol in the Presence of near-UV Illuminated Titanium Dioxide. Journal of Photochemistry and Photobiology A: Chemical 1992 (64) 231.

[54] Burns, R., Crittenden, J.C., Hand, D.W., Sutter, L.L., Salman, S.R., 1999. Effect of inorganic ions in heterogeneous photocatalysis. J. Environ. Eng. 125, 77-85.

Permissions

The contributors of this book come from diverse backgrounds, making this book a truly international effort. This book will bring forth new frontiers with its revolutionizing research information and detailed analysis of the nascent developments around the world.

We would like to thank M. Nageeb Rashed, for lending her expertise to make the book truly unique. She has played a crucial role in the development of this book. Without her invaluable contribution this book wouldn't have been possible. She has made vital efforts to compile up to date information on the varied aspects of this subject to make this book a valuable addition to the collection of many professionals and students.

This book was conceptualized with the vision of imparting up-to-date information and advanced data in this field. To ensure the same, a matchless editorial board was set up. Every individual on the board went through rigorous rounds of assessment to prove their worth. After which they invested a large part of their time researching and compiling the most relevant data for our readers. Conferences and sessions were held from time to time between the editorial board and the contributing authors to present the data in the most comprehensible form. The editorial team has worked tirelessly to provide valuable and valid information to help people across the globe.

Every chapter published in this book has been scrutinized by our experts. Their significance has been extensively debated. The topics covered herein carry significant findings which will fuel the growth of the discipline. They may even be implemented as practical applications or may be referred to as a beginning point for another development.

The editorial board has been involved in producing this book since its inception. They have spent rigorous hours researching and exploring the diverse topics which have resulted in the successful publishing of this book. They have passed on their knowledge of decades through this book. To expedite this challenging task, the publisher supported the team at every step. A small team of assistant editors was also appointed to further simplify the editing procedure and attain best results for the readers.

Our editorial team has been hand-picked from every corner of the world. Their multi-ethnicity adds dynamic inputs to the discussions which result in innovative outcomes. These outcomes are then further discussed with the researchers and contributors who give their valuable feedback and opinion regarding the same. The feedback is then collaborated with the researches and they are edited in a comprehensive manner to aid the understanding of the subject.

Apart from the editorial board, the designing team has also invested a significant amount of their time in understanding the subject and creating the most relevant covers. They scrutinized every image to scout for the most suitable representation of the subject and create an appropriate cover for the book.

The publishing team has been involved in this book since its early stages. They were actively engaged in every process, be it collecting the data, connecting with the contributors or procuring relevant information. The team has been an ardent support to the editorial, designing and production team. Their endless efforts to recruit the best for this project, has resulted in the accomplishment of this book. They are a veteran in the field of academics and their pool of knowledge is as vast as their experience in printing. Their expertise and guidance has proved useful at every step. Their uncompromising quality standards have made this book an exceptional effort. Their encouragement from time to time has been an inspiration for everyone.

The publisher and the editorial board hope that this book will prove to be a valuable piece of knowledge for researchers, students, practitioners and scholars across the globe.

List of Contributors

Zhengjun Zhang and Xian Zhang
Advanced Materials Laboratory, Department of Materials Science and Engineering, Tsinghua University, Beijing, P. R. China

Qin Zhou
Advanced Materials Laboratory, Department of Materials Science and Engineering, Tsinghua University, Beijing, P. R. China
Institute of nuclear and new energy technology, Tsinghua University, Beijing, P. R. China

Radim Vácha, Jan Skála, Jarmila Čechmánková and Viera Horváthová
Research Institute for Soil and Water Conservation, Prague, Czech Republic

Anna Białk-Bielińska, Jolanta Kumirska and Piotr Stepnowski
Department of Environmental Analysis, Faculty of Chemistry, University of Gdańsk, Gdańsk, Poland

Monia Renzi
Department of Environmental Science, Via Mattioli, University of Siena, Italy

Hongqi Wang, Shuyuan Liu and Shasha Du
College of Water Sciences, Beijing Normal University, Key Laboratory of Water and Sediment Sciences, Ministry of Education.Beijing, China

Claudio Cameselle and Susana Gouveia
Department of Chemical Engineering, University of Vigo, Building Fundicion, Vigo, Spain

Djamal Eddine Akretche and Boualem Belhadj
Laboratory of Hydrometallurgy and Inorganic Molecular Chemistry, Faculty of Chemistry, USTHB, BP 32, El- Alia, Bab Ezzouar, Algiers, Algeria

Mohamed Nageeb Rashed
Aswan Faculty of Science, Aswan University, Aswan, Egypt

Ana P.F. Santos, Douclasse C. de Castro, Jéssica A. Nogueira, Amilcar Machulek Jr, Silvio C. Oliveira, Valdir S. Ferreira Samuel, L. Oliveira Rodrigo P. Cavalcante, Fabio Gozzi, Dayana D. Ramos and Ana P.P. da Rosa
Center for Exact Sciences and Technology (CCET), Federal University of Mato Grosso do Sul-UFMS; Campo Grande, MS, Brazil

Marly E. Osugi
Institute of Chemistry, University of Brasília; Brasília, DF, Brazil

Frank H. Quina and Volnir O. Silva
Institute of Chemistry and NAP-PhotoTech - USP; University of São Paulo-USP; São Paulo, SP, Brazil

F. Dantas
Department of Chemical Engineering, Faculty of Chemistry, University of Barcelona, Barcelona, Spain

Gleison A. Casagrande
Faculty of Exact Sciences and Technology (FACET), Federal University of Grande Dourados- UFGD; Dourados, MS, Brazil

Fauze J. Anaissi
Department of Chemistry, State University of Centro-Oeste - UNICENTRO; Guarapuava, PR, Brazil

Muhammad Umar
School of Civil, Environmental and Chemical Engineering, RMIT University, Melbourne, Victoria, Australia

Hamidi Abdul Aziz
School of Civil Engineering, University Sains Malaysia, Engineering Campus, Nibong Tebal, Penang, Malaysia